THE NORMAL SERIES.

AN

INTELLECTUAL AND PRACTICAL

ARITHMETIC,

OR,

FIRST LESSONS

IN

ARITHMETICAL ANALYSIS.

INTENDED AS AN INTRODUCTION TO DODD'S ARITHMETIC.

BY JAMES L. ENOS.

NEW-YORK:
PRATT, WOODFORD & CO.,
NO. 4 CORTLAND-STREET.

1851.

PREFACE.

It has become a generally received truth, that *Intellectual Arithmetic* is among the first studies that should be presented to the juvenile mind. It is also a truth worthy of general reception, that the young grasp combinations of numbers with greater ease, when properly presented, than any other branch of science. This, together with the fact that whatever is taught a child at first should be UNCHANGING TRUTH, is ample evidence in favor of this branch as the FIRST STUDY OF EVERY CHILD.

The following Arithmetical Analysis is designed to present the subject in its most practical light—developing the *general principles* of numbers in a simple, yet strictly philosophic and logical manner.

The arrangement of the text, though in some respects different from any other work on this branch, is nevertheless believed to be a more *natural* arrangement, and at the same time to dictate a more simple analysis of the Elements of Arithmetic; a want of which has, unquestionably, led pupils to be content with a superficial knowledge of the principles of the science.

Multiplication is placed immediately after Addition, from the fact that it is only a different mode of applying the same general principle. Fractions are introduced immediately after Division, and the relation and relative size of one part to another are clearly elucidated. The questions throughout are of a practical character, or given with reference to securing the attention of the pupil, and holding it for a long time, on a single subject, thus strengthening the reasoning powers, and capacitating them for more powerful mental effort.

The mode of solving the questions is the one long used by the Author, and has received the commendation of many of the most distinguished teachers in the United States, and has been in substance adopted by them.

The first sections, and the first questions of each section, are formed of small numbers, the following ones rising in regular and systematic gradation, to greater degrees of complexity.

Thus the mind of the pupil is gradually led out and strengthened, and the foundation of a healthy and unchanging mental progression firmly implanted in the mind.

Exercises rationally performed serve a far better mode of awakening the faculties of children to invigorating effort, than by recourse to demonstrations of mere abstract propositions; and the present work has been prepared with direct reference to this important principle.

The *notes* are designed chiefly for the use of young teachers, or those who have never been called to teach this branch. These, together with the ANALYSIS of a large number of questions, it is believed will render the method of procedure so simple that teachers need have no fears in instructing a class, even though they have never pursued the study in school.

In submitting this little work to the consideration of his fellow laborers, the author would only add that, in its preparation, great pains has been taken, and most of the questions presented have been put to the practical test in the school room, before being incorporated into the work itself. He believes the *arrangement, system,* and *sufficient simplicity* of its pages will secure the largest portion of its success.

With this brief statement of the general plan of the work, it is most respectfully dedicated to the friends of education, trusting to its merits alone for adoption and use.

JAMES L. ENOS.

MADISON, WN., 1850.

ARITHMETIC

PART I.

CHAPTER I.

§ 1. *Note.* THE child is supposed to be able to count at least one hundred, and clearly understand what is denoted by the names used in expressing the several numbers. The age at which pupils may commence the study of this work will, of course, depend upon circumstances. Some children, who have a natural taste for mathematical studies, may commence at a very early age. Others will require to be much older.

Great care must be taken that all the examples contained in one section are fully understood before advancing to another.

Progress slowly and thoroughly—as this is the one sure way of securing a future rapid progress.

§ 2. One is represented by the character 1.

1. James had 1 book and his sister gave him 1 book more; how many had he then? 1 and 1 are how many?

2. If I have 1 book in my right hand and two in my left, how many have I in both hands? 1 and two are how many?

3. A boy had 1 marble and his brother gave him three more; how many had he then? How many are 1 and three?

4. In one room is 1 girl at play and in another four; how many are at play in both rooms? Why?

5. One man has 1 horse and another has five; how many horses do they both own? 1 and five are how many?

6. John gave 1 cent for an apple and seven cents for a primer; how many cents did he give for both?

7. William bought a marble for 1 cent and some pencils for eight cents; how many cents did he pay away?

8. James purchased a top for 1 penny and paid nine cents for other toys; how many did he pay? 1 and nine are how many?

9. A man distributed money among his children as follows; to James he gave three cents, to Harriet 1 cent, to John two cents, and to Mary 1 cent; how many cents did he give to all?

10. Two and 1 and three and four are how many? Two and 1 are how many? 1 and four? 1 and five? 1 and six? &c.

---

§ 3. Two is represented by the character 2.

1. If an apple cost 2 cents, and a fig cost 1 cent, how many cents will they both cost? 2 and 1 are how many?

2. James gave his sister three apples, and George gave her 2; how many did both together give her? Why?

3. William has four brothers and 2 sisters; how many brothers and sisters has he together? How many are four and 2?

4. Mary gave a blind man five cents and Samuel gave him 2 cents? how many cents did they both give him? Why?

5. Sarah recited six perfect lessons one day and 2 another; how many perfect lessons did she recite? How many are six and 2?

Note 2. Instead of presenting the whole of the ADDITION TABLE on a single page, it is presented in parts and followed by appropriate examples. Teachers cannot manifest too much caution in regard to passing over the pages of this work too rapidly. Pupils should be early taught that to insure a future RAPID PROGRESS, they must thoroughly learn every principle as they pass over it. After the following tables have been recited separately by each pupil, they may be given by the class in concert; for the purpose of rendering the exercise more pleasing and fastening them in the mind.

1	and	1	are	2.	1	and	six	are	seven.
1	"	2	"	three.	1	"	seven	"	eight.
1	"	three	"	four.	1	"	eight	"	nine.
1	"	four	"	five.	1	"	nine	"	ten.
1	"	five	"	six.	1	"	ten	"	eleven.

6. James, Martha and Mary had each 1 cent; how many had they together? Why?

7. John had 1 book and his mother gave him 2 books more; how many had he then? How many are 1 and 2?

8. William had six marbles in his hand and 1 in his pocket; how many marbles had he? How many are six and 1?

9. Henry gave Maria 1 book and her teacher gave her five; how many did she then have? Why?

10. Thomas gave seven cents for a knife and 1 cent for a pencil; how many cents did he spend? Why?

---

§ 4. Three is represented by the character 3.

1. If an orange cost 3 cents and an apple 1 cent, what will both together cost? How many are 3 and 1?

2. Mary gave 2 cents for a spool of thread and 3 cents for a skein of silk; how many cents did she spend? Why?

3. A man had 1 horse in his stable and 3 in each

of his two pastures; how many horses had he in all? Why?

4. How many are 2 bushels, 3 bushels and 1 bushel? Why?

5. A man had 3 dollars in one pocket, 3 dollars in his hand, and 2 dollars on his table; how many dollars had he? Why?

2	and	1	are	3.	2	and	six	are	eight.
2	"	2	"	four.	2	"	seven	"	nine.
2	"	3	"	five.	2	"	eight	"	ten.
2	"	four	"	six.	2	"	nine	"	eleven.
2	"	five	"	seven.	2	"	ten	"	twelve.

6. If you were to buy an ounce of figs for five cents and an orange for 2 cents, how many cents would you give for both?

7. If you had seven marbles in your pocket, and your playmate should give you 2 more, how many marbles would you have in all? Seven and 2 are how many?

8. Delia had eight dollars and her father gave her 2 more; how many did she then have? How many are eight and 2?

9. A man purchased a quantity of flour for nine dollars, and sold it for 2 dollars more than he gave for it; how many dollars did he sell it for?

10. A farmer bought ten pounds of coffee at one time and 2 at another; how many pounds did he buy in all? Why?

---

§ 5. Four is represented by the character 4.

1. James bought a slate for eight cents and a box of pencils for 3 cents; how many cents did he pay for both? How many are eight and 2?

2. George gave nine cents for an Arithmetic, and

3 cents for some quills; how many cents did he pay away?

3. Harriet gave a beggar ten cents and James gave him 2 cents more than Harriet; how many cents did James give him? Why?

4. Mary paid eleven cents for a writing book, 2 cents for an inkstand, and 1 cent for a pen; how many cents did she give for all?

5. John paid twelve cents for some raisins, 3 cents for some nuts, and then had 4 cents left; how many had he at first, and how many did he pay away?

3	and	1	are	4.	3	and	six	are	nine.
3	"	2	"	five.	3	"	sever.		ten.
3	"	3	"	six.	3	"	eight	"	eleven.
3	"	4	"	seven.	3	"	nine	"	twelve.
3	"	five	"	eight.	3	"	ten	"	thirteen.

6. A traveler paid at one tavern 3 dollars, 4 dollars at another, and 2 dollars at a third; how many dollars did he pay in all?

7. A farmer sold wheat in parcels as follows: to one man he sold 4 bushels, to another 3 bushels, and to another 2 bushels; how many bushels did he sell to all? How many are 4, 3, and 2?

8. Henrietta had five roses and her sister gave her 3 more; how many had she then? How many are five and 3?

9. Caroline gave six cents to a poor boy and Eliza gave him 3 more; how many did he receive from both girls? Why?

10. A farmer had seven cows in one pasture and 3 in another; how many had he in both? How many are seven and 3?

---

§ **6**. Five is represented by the character 5.

1. Frances gave 5 cents for a ball of twist, 3 cents

for a spool of thread, and 2 cents for some needles; how many cents did she spend? How many are 5, 3 and 2? 2, 3 and 5? 3, 2 and 5?

2. A farmer sold a firkin of butter for six dollars, some lard for 4 dollars, a quantity of poultry for 2 dollars, and a lamb for 1 dollar; how many dollars did he receive for all together? How many are six, 4, 2 and 1? 2, 1, six and 4?

3. A lady bought a yard of cambric for seven cents, a paper of pins for 3 cents, and a card of hooks and eyes for 4 cents; how many cents did she expend?

4. A man bought a barrel of cider for eight dollars, and sold it for two dollars more than he gave for it; how much did he sell it for?

5. A merchant sold at one time nine yards of calico, at another 5 yards, and 4 yards at another time; how many yards did he sell? Nine and 5 and 4 are how many?

4	and	1	are	5.	4	and	six	are	ten.
4	"	2	"	six.	4	"	seven	"	eleven.
4	"	3	"	seven.	4	"	eight	"	twelve.
4	"	4	"	eight.	4	"	nine	"	thirteen.
4	"	5	"	nine.	4	"	ten	"	fourteen.

6. A man bought a hundred weight of sugar for ten dollars, and a barrel of flour for 5 dollars; how much did he pay for the whole?

7. A man bought three barrels of flour for eleven dollars, and sold them again for 4 dollars more than he gave for them; how much did he sell them for?

8. A man has two notes due him, one of twelve dollars and the other of 4 dollars; how much is due him in all?

9. A man bought at one time nine tons of hay and at another 4 tons; how much did he buy in all? Why?

10. A farmer found on settling with the grocer with whom he dealt, that he had purchased coffee during the year as follows: at one time 5 pounds, at another *nine*, at another 3, at another 4, and at another 5; how many pounds had he bought in all?

---

§ 7. Six is represented by the character 6.

1. John bought an orange for 6 cents and some figs for 5 cents; what did both cost? How many are 6 and 5?

2. James lost 5 cents in the morning and 4 cents in the evening; how many did he lose in all?

3. A man paid eight dollars for a cow, 3 dollars for a hog, and 2 dollars for a sheep; how much did he pay for all? How many are eight, 3 and 2? 3, 2 and eight?

4. A class in Arithmetic contained ten girls and four boys; how many in the class?

5. Four boys went a fishing; the first caught eleven, the second 3, the third 2, and the fourth 1; how many did all together catch?

5	and	1	are	6.	5	and	6	are	eleven.
5	"	2	"	seven.	5	"	seven	"	twelve.
5	"	3	"	eight.	5	"	eight	"	thirteen.
5	"	4	"	nine.	5	"	nine	"	fourteen.
5	"	5	"	ten.	5	"	ten	"	fifteen.

6. In one township a drover bought cattle as follows: of one man he bought 6, of another 5, of another 3, and of another 1; how many did he purchase in all?

7. A market woman paid fifteen cents for a chicken, 5 cents for a piece of beef, 3 cents for a head of cabbage, and 2 cents for a turnip; how much money did she spend?

8. If, in a room, there are fourteen men, 5 women, 4 boys, and 3 girls, how many are there in the room? Why?

9. A boy paid thirteen cents for marbles, 4 cents for apples, and had 5 cents remaining; how many had he at first?

10. In a school there are twelve studying arithmetic, 3 learning geography, and 5 learning to spell; how many are in the school?

---

§ 8. Seven is represented by the character 7.

1. In one piece of cloth are 5 yards, and in another 7; how many yards in both?

2. A boy bought a fish-hook for 2 cents, a line for 4 cents, and a pole for 7 cents; how many cents did he give for the whole?

3. Joseph gave eight cents for a sled, and then gave 7 cents to have it painted; what was the whole cost of the sled?

4. James had 4 marbles, John gave him 3, George gave him 4, William gave him 5, and Thomas gave him 2; how many had he then?

5. A man sold five cows; gaining ten dollars on the first, 6 dollars on the second, nothing on the third, and one dollar on the fourth; how many dollars did he gain?

6	and	1	are	7.	6	and	6	are	twelve.
6	"	2	"	eight.	6	"	7	"	thirteen.
6	"	3	"	nine.	6	"	eight	"	fourteen.
6	"	4	"	ten.	6	"	nine	"	fifteen.
6	"	5	"	eleven.	6	"	ten	"	sixteen.

6. A man gave eleven dollars for a cart, 6 dollars for a plow, and 3 dollars for a harrow; how many dollars did they all cost him?

7. A school girl had 6 books; her brother gave her 5 more, and her mother gave her 7 more; how many had she then?

8. A merchant collected notes as follows; of one man sixteen dollars, of another 7, of another 6, and of another 5; how many dollars did he collect in all?

9. A drover bought sheep as follows: of one man fifteen, of another 7, of another 4, of another 3, and of another 6; how many did he buy in all?

10. Begin with 1 and count fifty-two, by adding 3 successively. Thus, 1 and 3 are 4, and 3 are 7, and 3 are ten, &c.

---

§ **9.** The number eight is represented by the character 8.

1. A boy having 8 apples, purchased 2 more; how many had he then? 8 and 2 are how many? 8 and 3? 8 and 4? 8 and 5?

2. James had three books, for which he paid as follows: for one he gave eleven cents, for another 8, and for another 3; how much did he pay for all? Why?

3. Commence with 2 and add 2 successively, until the number has reached one hundred.

4. Commence with 3 and count one hundred and three, by adding 5 successively.

5. The age of James is 8 years, of John 7 years, of Mary ten years, and of Jane 5 years; what is the sum of all their ages?

7	and	1	are	8.	7	and	6	are	thirteen.
7	"	2	"	nine.	7	"	7	"	fourteen.
7	"	3	"	ten.	7	"	8	"	fifteen.
7	"	4	"	eleven.	7	"	nine	"	sixteen.
7	"	5	"	twelve.	7	"	ten	"	seventeen.

6. Three boys talking together about their ages,

found that there was 3 years difference in the ages of the younger two, and the third was 2 years older than the elder of those two; the youngest was 7 years of age; what was the age of each boy, and what was the sum of all their ages?

7. A man bought a cow for *thirteen* dollars, and paid 5 dollars for keeping her; how much did she cost him?

8. A farmer took a load of produce to market, which he disposed of for the following sums of money; for his oats he received 7 dollars, for his potatoes 3 dollars, for his beans 5 dollars, and for his poultry 2 dollars; how much did his load come to?

9. Thomas bought a top for 7 cents, a knife for 6 cents, a comb for 4 cents, and a pen for 2 cents; what did they all cost him? 7 and 6 and 4 and 2 are how many?

10. A man paid twelve dollars for a barrel of pork, 3 dollars for a barrel of flour, and 7 dollars for a hundred weight of sugar; how much did he pay for the whole?

---

§ **10.** Nine is represented by the character 9.

1. James gave 8 cents for apples, 2 cents for a comb, and 1 cent for a stick of candy; how much did it all cost him? Why?

2. George was 8 years old and William 4; what was the sum of their ages?

3. Emeline gave 8 cents for some calico, 5 cents for some ribbon, and 3 cents for a spool of thread; how many cents did she pay for the whole?

4. Eliza paid 8 dollars for her cloak, and 6 dollars for a dress; how much did she pay for both?

5. Jane had 8 dollars in her purse, and her mother gave her 7 more; how many did she then have?

8 and	1	are	9.		8 and	6	are	fourteen.
8 "	2	"	ten.		8 "	7	"	fifteen.
8 "	3	"	eleven.		8 "	8	"	sixteen.
8 "	4	"	twelve.		8 "	9	"	seventeen.
8 "	5	"	thirteen.		8 "	ten	"	eighteen.

6. A man purchased of one of his neighbors 3 sheep for 8 dollars, of another 5 sheep for 4 dollars, and of a third 2 sheep for 5 dollars; how many sheep did he buy, and what did they all cost him?

7. In a school there are 9 writing, 8 studying grammar, 5 studying arithmetic, 3 studying geography, and 2 in algebra; how many in all?

8. A market-man received 6 dollars for butter, 7 dollars for cheese, and 9 dollars for poultry; how many dollars did he receive in all?

9. One man owes me 5 dollars, another 4 dollars, another 3 dollars, another 2 dollars, and another 9 dollars; how many dollars are due me in all?

10. A drover bought a number of hogs; of one man he bought 1, of another 2, of another 3, of another 8, of another 7, of another 6, of another 5, and of another 4; how many did he buy in all?

---

§ 11. Ten is represented by the characters 1 (one) and 0 (naught,) thus, 10.

1. Mary had 10 peaches, and her brother gave her 3 more; how many did she then have? 10 and 3 are how many?

2. James recited 9 perfect lessons in the forenoon and 5 in the afternoon; how many did he recite during the day?

3. John gave 10 cents to a beggar, and William gave him 6; how many cents did he receive from both boys?

9	and	1	are	10.	9	and	6	are	fifteen.
9	"	2	"	eleven.	9	"	7	"	sixteen.
9	"	3	"	twelve.	9	"	8	"	seventeen.
9	"	4	"	thirteen.	9	"	9	"	eighteen.
9	"	5	"	fourteen.	9	"	10	"	nineteen.

4. If an ox cost nineteen dollars and a sheep 2 dollars, what would be the cost of both?

5. A man bought five barrels of cider for 10 dollars, and nine bushels of apples for 5 dollars; how much did he pay for the whole?

6. A grocer sold in one day coffee as follows: to one man he sold 10 pounds, to another 4, to another 2, to another 3, to another 5, and to another 6; how many parcels did he sell in all?

7. John had 5 marbles, William 10, James 9, and George 6; how many had they all together?

8. If half a pound of tea cost eighteen cents, and six pounds of coffee cost 9 cents more, what would be the cost of six pounds of coffee?

9. How many are 10 dollars, 9 dollars, 8 dollars, and 7 dollars?

10. How many are 6 dollars, 5 dollars, 4 dollars, 3 dollars, 2 dollars, and 1 dollar?

---

11. How many are 2 and 1? 2 and 2? 2 and 3? 2 and 4? 2 and 5? 2 and 6? 2 and 7? 2 and 8? 2 and 9? 2 and 10?

12. How many are 3 and 2? 3 and 3? 3 and 4? 3 and 5? 3 and 6? 3 and 7? 3 and 8? 3 and 9? 3 and 10?

13. How many are 4 and 2? 4 and 3? 4 and 4? 4 and 5? 4 and 6? 4 and 7? 4 and 8? 4 and 9? 4 and 10?

14. How many are 5 and 2? 5 and 3? 5 and 4?

5 and 5? 5 and 6? 5 and 7? 5 and 8? 5 and 9? 5 and 10?

15. How many are 6 and 2? 6 and 3? 6 and 4? 6 and 5? 6 and 6? 6 and 7? 6 and 8? 6 and 9? 6 and 10?

16. How many are 7 and 2? 7 and 3? 7 and 4? 7 and 5? 7 and 6? 7 and 7? 7 and 8? 7 and 9? 7 and 10?

17. How many are 8 and 2? 8 and 3? 8 and 4? 8 and 5? 8 and 6? 8 and 7? 8 and 8? 8 and 9? 8 and 10?

18. How many are 9 and 2? 9 and 3? 9 and 4? 9 and 5? 9 and 6? 9 and 7? 9 and 8? 9 and 9? 9 and 10?

19. How many are 10 and 2? 10 and 3? 10 and 4? 10 and 5? 10 and 6? 10 and 7? 10 and 8? 10 and 9? 10 and 10?

20. How many are 1 and 2, and 3, and 4, and 5, and 6, and 7, and 8, and 9, and 10?

---

21. A bookseller sold in one day 10 arithmetics, 9 geographies, 8 grammars, and 5 spelling books; how many books did he sell in the day?

22. Emma went to make some purchases; for a silk dress she paid eleven dollars, 6 dollars for a satin bonnet, and 5 dollars for a number of smaller articles; how much money did she pay?

23. George bought a wagon for eighteen dollars; he gave 6 dollars to have it repaired, and 3 dollars to have it painted, and then sold it for 7 dollars more than it had cost him; how many dollars did he receive for it?

24. A grocer sold to one man fifteen pounds of sugar, to another 5 pounds, to another 8 pounds, to another 2 pounds, and to another 10 pounds; how many pounds of sugar did he sell?

25. One boy pays 10 dollars a term for his tuition, another pays 3 dollars, another 7 dollars, another 5 dollars, and another 4 dollars; how much do all together pay?

26. A man gave fourteen dollars to one of his children, 3 to another, 5 to another, 10 to another, 9 to another, and to another he gave 7 dollars, when he found he had just 8 dollars remaining; what number of dollars had he at first?

CHAPTER II.

§ **12.**—1. James had one apple for which he gave 1 cent; what would be the cost of 2 apples at the same rate?

2. How many is 2 times 1?

3. Harriet paid 2 cents for one peach, and 2 cents for another; how much did she pay for both peaches? 2 times 2 is how many?

4. Mary bought 3 yards of tape at 2 cents a yard; what did it cost her? 2 times 3 is how many?

5. John bought 2 pencils at 4 cents a piece; what did they come to?

2	times	1	is	2.	2	times	6	is	twelve.
2	"	2	"	4.	2	"	7	"	fourteen.
2	"	3	"	6.	2	"	8	"	sixteen.
2	"	4	"	8.	2	"	9	"	eighteen.
2	"	5	"	10.	2	"	10	"	twenty.

Note. The manner of representing numbers by figures has been explained as far as 10. Those from 10 to five hundred are written as follows:

Eleven is written		11.	Twenty-eight is written		28.
Twelve	"	12.	Twenty-nine	"	29.
Thirteen	"	13.	Thirty	"	30.
Fourteen	"	14.	Thirty-one	"	31.
Fifteen	"	15.	Forty	"	40.
Sixteen	"	16.	Fifty	"	50.
Seventeen	"	17.	Sixty	"	60.
Eighteen	"	18.	Seventy	"	70.
Nineteen	"	19.	Eighty	"	80.
Twenty	"	20.	Ninety	"	90.
Twenty-one	"	21.	One hundred	"	100.
Twenty-two	"	22.	1 hundred and 1	"	101.
Twenty-three	"	23.	1 hundred and 2	"	102.
Twenty-four	"	24.	Two hundred	"	200.
Twenty-five	"	25.	Three hundred	"	300.
Twenty-six	"	26.	Four hundred	"	400.
Twenty-seven	"	27.	Five hundred		500.

6. What would be the cost of 4 spools of thread at 2 cents a spool?

7. What would be the cost of 5 yards of tape at 2 cents a yard?

8. If one pound of sugar cost 6 cents, what will 2 pounds cost? 2 times 6 is how many?

Solution.—If one pound of sugar cost 6 cents, 2 pounds will cost 2 times 6 cents; 2 times 6 cents are twelve cents. Therefore, 2 pounds of sugar at 6 cents a pound will cost 12 cents.

9. If one barrel of pork cost 7 dollars, what will 2 barrels cost?

10. What will be the cost of 2 barrels of flour at 8 dollars a barrel?

11. What will be the cost of 2 quires of paper at 9 cents a quire?

12. What will be the cost of 2 coats at 10 dollars a piece?

§ **13.**—1. If one bushel of wheat cost 1 dollar, what will be the cost of 3 bushels? 3 times 1 is how many?

3 times 1 is 3.	3 times 5 is 15.	3 times 9 is 27.
3 " 2 " 6.	3 " 6 " 18.	3 " 10 " 30.
3 " 3 " 9.	3 " 7 " 21.	3 " 11 " 33.
3 " 4 " 12.	3 " 8 " 24.	3 " 12 " 36.

2. What cost 2 peaches, at 3 cents a piece?

3. What must you pay for 4 oranges at 2 cents a piece?

4. What cost 5 barrels of cider at 2 dollars a barrel?

5. What will 6 quarts of berries cost, at 2 cents a quart?

6. At 2 cents a piece, what will 7 lead pencils cost?

7. A man gave 2 boys 8 apples a piece; how many did he give to both?

8. If it take 9 yards of cloth to make a cloak, how many yards will be required for 2 cloaks?

9. What cost 10 yards of cloth, at 2 dollars a yard?

10. If it take 2 bushels of wheat to sow an acre of ground, how many bushels will it take to sow 11 acres?

11. If my board cost me 2 dollars a week, what will it amount to in 12 weeks?

12. James bought 3 apples at 2 cents a piece, and 2 combs at 5 cents a piece; how many cents did he pay for the whole?

---

13. What must I pay for 3 oranges, at 3 cents a piece?

Solution.—If I pay 3 cents for 1 orange, I must pay 3 times 3 cents for 3 oranges; 3 times 3 cents are 9 cents. Therefore, for 3 oranges at 3 cents a piece, I must pay 9 cents.

14. If one barrel of flour cost 3 dollars, what will 4 barrels cost?

4 times 1 is 4.	4 times 5 is 20.	4 times 9 is 36.
4 " 2 " 8.	4 " 6 " 24.	4 " 10 " 40.
4 " 3 " 12.	4 " 7 " 28.	4 " 11 " 44.
4 " 4 " 16.	4 " 8 " 32.	4 " 12 " 48.

15. What will four bushels of wheat cost at 1 dollar a bushel?

16. At 2 cents a piece what will 4 lemons cost?

17. Jane bought 3 spools of thread at 4 cents a piece; how much did they all cost her?

18. What cost 4 yards of cloth at 4 dollars a yard?

19. In one month there are four weeks; how many weeks are there in 5 months?

20. In one bushel there are 4 pecks; how many pecks are there in 5 bushels?—in 3 bushels?—in 7 bushels?

21. What cost 6 lemons at 4 cents a piece?

22. If a man can travel 4 miles in one hour, how far can he travel in 7 hours?—in 4 hours?—in 9 hours?

23. In one gallon there are 4 quarts; how many quarts are there in 8 gallons?—in 5 gallons?—in 10 gallons?

24. At 3 cents a piece, what will 12 primers cost?—11?—5?

25. If you can buy 4 peaches for one cent, how many can you buy for 9 cents?—for 7 cents?—for 11 cents?

26. A gardener sold 12 barrels of quinces, at 4 dollars a barrel; how many dollars did they amount to?

---

27. At 5 dollars a yard, what would 2 yards of cloth cost?

Solution.—If one yard cost 5 dollars, 2 yards will cost 2 times 5 dollars; 2 times 5 dollars are 10 dollars. Therefore, 2 yards of cloth at 5 dollars a yard, would cost 10 dollars.

5 times 1 is 5.	5 times 5 is 25.	5 times 9 is 45.
5 " 2 " 10.	5 " 6 " 30.	5 " 10 " 50.
5 " 3 " 15.	5 " 7 " 35.	5 " 11 " 55.
5 " 4 " 20.	5 " 8 " 40.	5 " 12 " 60.

28. In one quart there are 2 pints; how many pints are there in 5 quarts?—in 2?—in 3?—in 4?

29. How many quarts are there in 5 gallons?—in 4?—in 2?

30. If there are 3 feet in one yard, how many feet are there in 5 yards?

31. How many feet are there in 11 yards?—in 12?—in 4?

32. What is the price of 10 yards of cloth, at 5 dollars a yard?

33. What cost 9 pounds of sugar, at 5 cents a pound?

34. In one fathom there are 6 feet; how many feet are there in 5 fathoms?

35. When flour is 5 dollars a barrel, what will be the cost of 8 barrels?—of 9?—of 10?—of 11?—of 12?

36. How many feet are there in 5 fathoms and 5 feet?

37. What will 5 barrels of pork cost, at 12 dollars a barrel?—at 11?—at 10?—at 9?—at 8?—at 7?—at 6?

---

38. What would be the cost of 6 bushels of wheat at 2 dollars a bushel?

6 times 1 is 6.	6 times 5 is 30.	6 times 9 is 54
6 " 2 " 12.	6 " 6 " 36.	6 " 10 " 60
6 " 3 " 18.	6 " 7 " 42.	6 " 11 " 66.
6 " 4 " 24.	6 " 8 " 48.	6 " 12 " 72.

39. If 2 men can do a piece of work in 6 days, in what time can 1 man perform the same work?

40. What cost 6 oranges at 3 cents a piece?

41. If a stage go 6 miles an hour, how far will it go in 4 hours? in 3 hours? in 7 hours? in 9 hours?

42. At 6 cents a pound, what will 5 pounds of butter cost? 6 pounds? 8 pounds? 10 pounds?

43. In one peck there are 8 quarts; how many quarts are there in 6 pecks? in 2 pecks? in 4 pecks?

44. James paid 6 cents for a slate; what would he have to pay for 9 slates at the same rate?

45. When flour is 6 dollars a barrel, what must I pay for 10 barrels? for 6? 11? 5? 9? 4? 3? 8?

46. At 11 cents a quart, what will 6 quarts of cherries cost? 5 quarts? 4 quarts? 3 quarts?

47. If one yard of cloth cost 12 dollars, what would be the cost of 6 yards? of 2 yards? of 5 yards?

48. A man sold 7 sheep for 3 dollars apiece, and 5 cows for one hundred dollars; how much did he receive for the whole?

49. At 3 dollars a cord, what will 11 cords of wood come to? 7 cords? 9 cords? 10 cords?

50. A jeweler sold 5 watches for 12 dollars each, how much did they all come to?

---

51. If 2 men can do a piece of work in 7 days, how many days will it take one man to do it?

SOLUTION.—If it take 2 men 7 days, it will take 1 man 2 times 7 days; 2 times 7 days are 14 days. Therefore, it would take 1 man 14 days to do the work.

7	times	1	is	7.	7	times	5	is	35.	7	times	9	is	63.
7	"	2	"	14.	7	"	6	"	42.	7	"	10	"	70.
7	"	3	"	21.	7	"	7	"	49.	7	"	11	"	77.
7	"	4	"	28.	7	"	8	"	56.	7	"	12	"	84.

52. If 3 men can perform a job of work in 7 days, how many days will it require for 1 to do it?

53. If a man can earn 7 dollars in one week, how many dollars can he earn in 4 weeks? in 2 weeks?

54. If the interest of 1 dollar is 7 cents a year, what will be the interest of 5 dollars for the same time? of 4 dollars? of 6 dollars? of 2 dollars? of 7 dollars?

55. Harriet had 7 rose bushes, and one morning she found 6 roses upon each; how many were there upon all?

56. At 7 cents a pound, what cost 7 pounds of lard?

57. If 1 yard of cloth cost 8 cents, what is the cost of 7 yards at the same rate?

58. What will 9 bushels of peaches cost at 7 shillings a bushel?

59. There are 10 rows of trees in an orchard, and 7 trees in a row; how many trees are there in the orchard?

60. At 7 dollars a week, what will 11 weeks board come to?

61. In one week there are 7 days, how many days are there in 12 weeks? in 9 weeks? in 6 weeks? in 3 weeks?

62. A merchant sold 7 yards of calico at 12 cents a yard, how much did he receive for it?

63. How much can a man earn in 7 months at 9 dollars a month? at 10 dollars? at 7 dollars?

---

64. In New York 8 shillings make 1 dollar; how many New York shillings in 2 dollars?

8 times	1	is	8.	8 times	5	is	40.	8 times	9	is	72.
8 "	2	"	16.	8 "	6	"	48.	8 "	10	"	80.
8 "	3	"	24.	8 "	7	"	56.	8 "	11	"	88.
8 "	4	"	32.	8 "	8	"	64.	8 "	12	"	96.

65. If a boat sail 8 miles in an hour, how far will it sail in 3 hours? in 2 hours? in 5 hours?

66. If in 1 month there are 4 weeks, how many weeks are there in 8 months? in 4 months? in 6 months?

67. What will 5 pounds of butter cost at 8 cents a pound? at 9 cents a pound? at 5 cents?

68. How many New York shillings are there in 6 dollars?

69. What is the value of 8 yards of cloth at 7 dollars a yard? at 5 dollars? at 3 dollars? at 9 dollars?

70. What will 8 barrels of vinegar cost at 8 dollars a barrel? at 2 dollars? at 4 dollars? at 6 dollars?

71. What is the value of 6 barrels of cider at 2 dollars a barrel?

72. At 3 dollars a yard, what would 7 yards of cloth cost?

73. In 1 penny there are 4 farthings; how many farthings are there in 12 pence? in 10 pence? in 8 pence?

74. Two men start from the same place and travel different ways; one at the rate of 3 miles in an hour, and the other 5 miles in an hour; how far apart will they be at the end of the first hour? How far at the end of 2 hours? 4 hours? 5 hours? 8 hours? 9 hours?

75. A boy bought 8 lemons at 3 cents a piece; 2 apples at 2 cents a piece; and 4 pears at 1 cent a piece; how much did the whole come to?

76. A ship loaded with emigrants, underwent an examination; when it was found she had on board 5 passengers from Spain; three times that number who were from Ireland; 3 were from Scotland, and as many from Germany as all beside; how many were from Germany, and how many passengers were there in all?

77. If 1 bale of cotton cost 9 dollars, what will be the cost of 2 bales ?

9	times	1	is	9.	9	times	5	is	45.	9	times	9	is	81.
9	"	2	"	18.	9	"	6	"	54.	9	"	10	"	90.
9	"	3	"	27.	9	"	7	"	63.	9	"	11	"	99.
9	"	4	"	36.	9	"	8	"	72.	9	"	12	"	108.

78. In 1 yard there are 3 feet; how many feet are there in 9 yards ? in 2 yards ? in 8 yards ? in 11 yards ?

79. How many feet are there in 9 yards, and 2 feet ?

80. If 1 pint of oil cost 9 cents, what will 1 quart cost ?

81. If 1 quart of beer cost 7 cents, what will be the cost of one gallon ?

82. If a man can build 9 rods of plank road in 1 day, how far could he build in a week ?

83. If a stage run 7 miles in an hour, how far will it run in 9 hours ? in 4 hours ? in 2 hours ? in 11 hours ?

84. A traveller met 9 beggars and gave each of them 8 cents ; how many cents did he give to all ?

85. What cost 9 bushels of corn at 9 cents a bushel ?

86. What cost 10 coats at 9 dollars each ?

87. A drover bought 9 cows and paid 11 dollars a piece for them ; how much did they all cost him ?

88. If a man can earn 12 dollars a month, how much can he earn in 9 months ?

89. If 11 men can do a piece of work in 9 days, how long will it take 1 man to do it ?

90. Two boats pass the same point on a river at the same time ; the one going north moves at the rate of 11 miles an hour ; and the one going south moves 12 miles an hour ; how far will they be apart in one hour ; in 2 hours ?

91. If one apple-tree bear 11 bushels of apples, how many bushels will 2 trees bear? 3 trees? &c.

10 times 1 is 10.	10 times 5 is 50.	10 times 9 is 90.
10 " 2 " 20.	10 " 6 " 60.	10 " 10 " 100.
10 " 3 " 30.	10 " 7 " 70.	10 " 11 " 110.
10 " 4 " 40.	10 " 8 " 80.	10 " 12 " 120.

92. If 1 barrel of flour cost 3 dollars, how much will 10 barrels cost?

93. In one dime there are 10 cents; how many cents are there in 4 dimes?

94. There are 10 dimes in one dollar; how many are there in 5 dollars? in 6 dollars? in 4 dollars?

95. If there are 10 dollars in one eagle, how many dollars are there in 6 eagles? in 7? in 8? in 9?

96. If you walk 10 miles in a day, how far can you walk in 7 days?

97. If you work 8 hours each day, how many hours would you work in 10 days?

98. How much will 9 yards of cloth come to at 10 shillings a yard?

99. A man purchased 11 acres of land for 10 dollars an acre, how much did the whole cost him?

100. A meteor in passing through the heavens is observed moving at the rate of 12 miles a minute; how far will it move in 10 minutes? in 4 minutes? in 8 minutes?

---

101. What will 2 pounds of butter cost at 11 cents a pound?

11 times 1 is 11.	11 times 5 is 55.	11 times 9 is 99.
11 " 2 " 22.	11 " 6 " 66.	11 " 10 " 110.
11 " 3 " 33.	11 " 7 " 77.	11 " 11 " 121.
11 " 4 " 44.	11 " 8 " 88.	11 " 12 " 132.

102. In one quarter there are 11 weeks; how many weeks in 3 quarters? in 4? in 5? in 2?

103. At 4 shillings a gallon, what would 11 gallons of molasses cost?

104. If a bookseller gain 5 cents on each book he sells, how much will he gain on the sale of 11 books?

105. A man bought 11 sheep, paying at the rate of 6 dollars a head; how much did he pay for the whole?

106. In a certain school there were 11 seats and 7 pupils on a seat; how many pupils were there in the school?

107. If the interest on 1 dollar for 1 month is 3 cents, how much will be the interest on 3 dollars for 11 months?

108. What will 9 months' work come to, at 11 dollars a month?

109. When hay is 10 dollars a ton, what will 11 tons come to?

110. If 11 cords of wood will last one family a year, how many cords will last 11 families the same time? 3 families?

111. A man bought 11 pounds of candles, at 12 cents a pound; how much did he pay for them?

112. 11 times 12 is how many? 11 times 7? 11 times 11? 11 times 5? &c.

---

113. If 1 barrel of cider last 3 persons 12 weeks, how many weeks will it serve 1 person?

12 times 1 is 12.	12 times 5 is 60.	12 times 9 is 108.
12 " 2 " 24.	12 " 6 " 72.	12 " 10 " 120.
12 " 3 " 36.	12 " 7 " 84.	12 " 11 " 132.
12 " 4 " 48.	12 " 8 " 96.	12 " 12 " 144.

114. At 2 dollars a gallon, what would 12 gallons of oil cost?

115. A laborer received 3 dollars a day for his work; how much would his wages amount to in 2 weeks, allowing 6 working days to the week?

116. At 12 cents an hour, how much must you pay for a horse and carriage to drive 4 hours?

117. Ten dimes make 1 dollar; how many dimes in 12 dollars?

118. If it take 6 knots of yarn to knit one pair of stockings, how many knots will it require to knit 12 pairs?

119. How much are 12 trunks worth, at 7 dollars a piece?

120. If 1 cow is worth 11 dollars, what are 12 cows worth?

121. When wheat is 7 dollars a hundred weight, what will be the cost of 12 hundred weight?

122. If 1 coat cost 5 dollars, what will 12 coats cost?

123. What will 1 bushel of wheat come to at 12 cents a peck?

124. If 1 bar of lead is worth 7 dollars, what would be the cost of 12 bars at the same rate?

125. At 3 dollars a week, what will 11 weeks' board come to?

126. If a man can earn 6 dollars in 1 week, how many dollars can he earn in 9 weeks? in 8 weeks? in 7 weeks?

127. A. purchased 3 hogs at 3 dollars each; 4 sheep at 4 dollars each; and 7 calves at 7 dollars each; how much did he give for the whole?

128. William, George and Mary had each 4 dollars, and James had as many again as all of them; how many dollars had James?

129. A farmer gave 6 dollars for a yoke, and 5 times as much for a pair of oxen; how much did he give for the whole?

130. What cost 11 pairs of boots at 7 dollars a pair?

131. What cost 7 pairs of shoes at 3 dollars a pair?

CHAPTER III.

§ 14.—1. James had 2 books, and gave 1 of them to Delia; how many had he left? Why?

2. Harriet had 3 roses, and gave 1 to Sarah and 1 to Mary; how many had she left? Why?

3. A boy had 4 marbles, but he lost 2 of them; how many had he remaining?

Solution.—He had the difference between 4 marbles and 2 marbles; 2 marbles from 4 marbles leaves 2 marbles. Therefore he had 2 marbles remaining.

2	from	2	leaves	0	3	from	9	leaves	6
2	"	3	"	1	3	"	10	"	7
2	"	4	"	2	3	"	11	"	8
2	"	5	"	3	3	"	12	"	9
2	"	6	"	4	3	"	13	"	10
2	"	7	"	5					
2	"	8	"	6	4	from	4	leaves	0
2	"	9	"	7	4	"	5	"	1
2	"	10	"	8	4	"	6	"	2
2	"	11	"	9	4	"	7	"	3
2	"	12	"	10	4	"	8	"	4
					4	"	9	"	5
3	from	3	leaves	0	4	"	10	"	6
3	"	4	"	1	4	"	11	"	7
3	"	5	"	2	4	"	12	"	8
3	"	6	"	3	4	"	13	"	9
3	"	7	"	4	4	"	14	"	10
3	"	8	"	5					

Note.—The teacher should illustrate by means of visible objects the process of taking one number from another. Books, the members of the class, beans, pieces of wood, or any objects that may be convenient. Let the pupil frequently recite the tables until they become firmly fixed in the mind.

4. A man bought seven pounds of coffee, and lost 5 pounds of it on his way home; how much had he remaining?

5. William had 12 pears, and gave his mother all but four of them; how many did he give her?

6. Emma had 13 cents, and she gave 5 cents for some thread; how much had she left?

7. A man had 20 sheep in a pasture, and lost 9 of them; how many had he remaining? How many more than he lost?

8. George bought a horse for 25 dollars, and paid 4 dollars for his keeping, and one dollar for having him shod, and then sold him for 30 dollars; did he gain or lose by the bargain, and how much?

9. A, B, and C, talking of their money, found that A had 15 dollars, B had 5 dollars less, and C had as much as A and B both, lacking 3 dollars; how much money had C, and how much had they all together?

10. 14 less 4 are how many? 13 less 3? 12 less 2? 11 less 4? 9 less 4? 7 less 3?

---

11. 13 less 3 are how many? 13 less 4? 13 less 2? 10 less 5? 9 less 4? 7 less 5? 6 less 2? 12 less 4? 12 less 3? 11 less 4?

12. A man had 14 dollars; he paid one man 3 dollars, and another 8 dollars; how many dollars did he pay away, and how many had he remaining?

Note.—The teacher can continue similar questions, of his own construction to any extent. They should be continued until the pupil is perfectly familiar with the principle involved.

13. A man had 15 dollars, and bought 40 pounds of sugar for 6 dollars; how much money had he left?

14. If a cow cost 16 dollars, and a colt 5 dollars, how much more does the cow cost than the colt?

5 from 5 leaves 0.
5 " 6 " 1.
5 " 7 " 2.
5 " 8 " 3.
5 " 9 " 4.
5 " 10 " 5.
5 " 11 " 6.
5 " 12 " 7.
5 " 13 " 8.
5 " 14 " 9.
5 " 15 " 10.

6 from 6 leaves 0.
6 " 7 " 1.
6 " 8 " 2.
6 " 9 " 3.
6 " 10 " 4.
6 " 11 " 5.
6 " 12 " 6.
6 " 13 " 7.
6 " 14 " 8.
6 " 15 " 9.
6 " 16 " 10.

7 from 7 leaves 0.
7 " 8 " 1.
7 " 9 " 2.
7 " 10 " 3.
7 " 11 " 4.
7 " 12 " 5.
7 " 13 " 6.
7 " 14 " 7.
7 " 15 " 8.
7 " 16 " 9.
7 " 17 " 10.

15. A man bought a tub of lard for 13 dollars, but it being damaged, he was obliged to sell it for 7 dollars less than it cost him ; what did he sell it for ?

16. A man bought 3 tons of hay for 15 dollars, and in pay for it he gave 3 bushels of wheat at 4 dollars, and 16 pounds of butter at 2 dollars, and the rest in money ; how much money did he pay ?

17. A man bought 10 barrels of cider for 14 dollars, and sold one half of it for the same that he gave for the whole ; if he sells the remainder at the same rate, how much will he gain by the sale ?

18. A farmer bought a sheep for 11 dollars, and to pay for it he gave seven bushels of corn worth 3 dollars, and the rest in money ; how much money did he pay ?

19. A man bought a sleigh for 7 dollars and sold it for 13 dollars ; how much did he gain by the bargain ?

20. A painter bought 12 gallons of oil, and after using a quantity of it found he had 5 gallons remaining; how much had he used?

21. Ellen had 18 cents, but she has spent 8 of them; how many has she remaining?

8	from	8	leaves	0.	9	from	15	leaves	6.
8	"	9	"	1.	9	"	16	"	7.
8	"	10	"	2.	9	"	17	"	8.
8	"	11	"	3.	9	"	18	"	9.
8	"	12	"	4.	9	"	19	"	10.
8	"	13	"	5.					
8	"	14	"	6.	10	from	10	leaves	0.
8	"	15	"	7.	10	"	11	"	1.
8	"	16	"	8.	10	"	12	"	2.
8	"	17	"	9.	10	"	13	"	3.
8	"	18	"	10.	10	"	14	"	4.
					10	"	15	"	5.
9	from	9	leaves	0.	10	"	16	"	6.
9	"	10	"	1.	10	"	17	"	7.
9	"	11	"	2.	10	"	18	"	8.
9	"	12	"	3.	10	"	19	"	9.
9	"	13	"	4.	10	"	20	"	10.
9	"	14	"	5.					

22. A grocer had 13 chests of tea, and sold 8 of them; how many had he left?

23. Sarah had 14 doves; she sold 9 of them and kept the remainder; how many did she keep?

24. A man sold a cow for 20 dollars, which was 10 dollars more than she cost him; how much did he give for her?

25. What would be the difference in the cost of 9 yards of cloth at 2 dollars a yard, and 4 tons of hay at 2 dollars a ton?

26. A man having 19 cords of wood, sold 3 cords to one man, and 7 to another; how many cords had he left?

Solution.—He had left the difference between the sum of 3 cords and 7 cords, and 19 cords. 3 cords and 7 cords are 10 cords; 10 cords from 19 cords leaves 9 cords. Therefore, he had 9 cords left.

27. In one field there are 15 acres of land, and in another 9 acres; how many more acres in one field than in the other? How many in both together?

28. 8 and 6 and 2 and 4 and 5, less 3, are how many?

Note.—This, and the following 10 examples should be solved as follows:—8 and 6 are 14, and 2 are 16, and 4 are 20, and 5 are 25. 3 from 25 leaves 22.

---

29. 7 and 5 and 3 and 5 and 2 and 1 and 4 and 8 and 6, less 8, are how many?

30. 8 and 6 and 4 and 6 and 3 and 2 and 5 and 9 and 7, less 9, are how many?

31. 9 and 7 and 5 and 7 and 4 and 3 and 6 and 10 and 8, less 10, are how many?

32. 2 and 8 and 6 and 8 and 5 and 4 and 7 and 11 and 9, less 11, are how many?

33. 3 and 9 and 6 and 9 and 5 and 7 and 4 and 1 and 10, less 1, are how many?

34. 4 and 1 and 7 and 2 and 6 and 8 and 5 and 2 and 1, less 2, are how many?

35. 5 and 2 and 8 and 3 and 7 and 9 and 3 and 4 and 2, less 3, are how many?

36. 6 and 3 and 9 and 4 and 8 and 10 and 4 and 5, less 4 and 3, are how many?

37. 7 and 4 and 10 and 5 and 9 and 11 and 5 and 6, less 5 and 2 are how many?

38. 8 and 5 and 11 and 6 and 10 and 12 and 6, less 7 and 6 and 3, are how many?

CHAPTER IV.

§ 15.—1. How many apples, at 1 cent a piece, can you buy for 2 cents?

2. At 2 cents a piece, how many figs can you buy for 4 cents?

3. If pins are 3 cents a paper, how many papers can you buy for 6 cents?

4. How many yards of tape, at 1 cent a yard, can you buy for 7 cents? for 5 cents? for 9 cents? for 11 cents? for 12 cents?

5. If a man can earn 2 dollars in 1 day, how many days will it take him to earn 8 dollars?

Solution.—If he earn 2 dollars each day, it will take him as many days to earn 8 dollars, as the number of times that 2 dollars are contained in 8 dollars. 2 dollars in 8 dollars, 4 times. Therefore, at 2 dollars a day, he must work 4 days to earn 8 dollars.

Note.—The following tables should be thoroughly learned, and the relation they sustain to the others, clearly explained.

2	in	2,	1 (once)	3	in	3,	1 (once)	4	in	4,	1 (once)
2	"	4,	2 times.	3	"	6,	2 times.	4	"	8,	2 times.
2	"	6,	3 "	3	"	9,	3 "	4	"	12,	3 "
2	"	8,	4 "	3	"	12,	4 "	4	"	16,	4 "
2	"	10,	5 "	3	"	15,	5 "	4	"	20,	5 "
2	"	12,	6 "	3	"	18,	6 "	4	"	24,	6 "
2	"	14,	7 "	3	"	21,	7 "	4	"	28,	7 "
2	"	16,	8 "	3	"	24,	8 "	4	"	32,	8 "
2	"	18,	9 "	3	"	27,	9 "	4	"	36,	9 "
2	"	20,	10 "	3	"	30,	10 "	4	"	40,	10 "
2	"	22,	11 "	3	"	33,	11 "	4	"	44,	11 "
2	"	24,	12 "	3	"	36,	12 "	4	"	48,	12 "

6. A man bought 10 dollars worth of shoes, at 2 dollars a pair; how many pair did he buy?

7. If 1 fish cost 2 cents, how many can be bought for 10 cents?

8. A man had 8 dollars to pay for wheat, at 2 dollars a bushel; how many bushels could he buy?

9. In 4 gallons there are 16 quarts; how many quarts in 1 gallon?

10. A tailor received 21 dollars for making 3 coats; how much was that for each coat?

11. A man bought 2 plows for 12 dollars; what was that a piece?

12. In 1 bushel are 4 pecks; how many pecks in 40 bushels?

13. A lady paid 36 cents for 4 yards of cambric, how much was that a yard?

14. Alice picked from her rose bushes 33 roses, which she distributed equally among her 3 schoolmates; how many did each receive?

15. A miller sold 2 barrels of flour for 12 dollars; what was that a barrel?

16. A brewer sold 4 barrels of beer for 24 dollars; what was that a barrel?

17. In a school containing 48 girls, are 4 classes in reading, and each class contains the same number; how many in a class?

18. If you had 24 cents, how many books could you buy at 4 cents a piece?

19. If a man can travel 3 miles in an hour, how long will it take him to travel 27 miles?

20. How many sheep, at 4 dollars a piece, can you buy for 16 dollars?

21. A man purchased a quantity of salt, paying at the rate of 4 dollars a barrel; he expended 36 dollars; how many barrels did he buy?

22. If a man can travel 4 miles in an hour, how long will it take him to travel 44 miles?

23. If you had 9 three dollar bills, how many yards of cloth could you buy at 3 dollars a yard?

24. In an orchard are 24 trees and 4 trees in a row; how many rows in the orchard?

25. If 1 man can do a piece of work in 48 days, in what time can 4 men do the same work?

26. At 3 dollars an acre, how many acres of land can you buy for 27 dollars?

27. If a stage go 4 miles an hour, how long would it be in going 28 miles? 16 miles? 12? 8? 20?

28. How long would it require for a man to go 10 miles, if he travel 5 miles an hour?

5 in 5,	1 (once)	6 in 6,	1 (once)	7 in 7,	1 (once)
5 " 10,	2 times.	6 " 12,	2 times.	7 " 14,	2 times.
5 " 15,	3 "	6 " 18,	3 "	7 " 21,	3 "
5 " 20,	4 "	6 " 24,	4 "	7 " 28,	4 "
5 " 25,	5 "	6 " 30,	5 "	7 " 35,	5 "
5 " 30,	6 "	6 " 36,	6 "	7 " 42,	6 "
5 " 35,	7 "	6 " 42,	7 "	7 " 49,	7 "
5 " 40,	8 "	6 " 48,	8 "	7 " 56,	8 "
5 " 45,	9 "	6 " 54,	9 "	7 " 63,	9 "
5 " 50,	10 "	6 " 60,	10 "	7 " 70,	10 "
5 " 55,	11 "	6 " 66,	11 "	7 " 77,	11 "
5 " 60,	12 "	6 " 72,	12 "	7 " 84,	12 "

29. A man had 15 dollars in three equal sized bills; what was the value of each?

30. At 5 dollars a yard, how many yards of carpeting can be bought for 20 dollars?

31. If a canal boat move 5 miles in an hour, how long will it be in going 25 miles?

32. From Chicago to Elgin is 30 miles; how long would it take a man to go the distance if he travel 5 miles an hour?

4

33. At 5 dollars a bushel, how much clover-seed can be bought for 35 dollars? for 40 dollars?

34. A man paid 45 dollars for 5 cows; how much were they a piece?

35. A man divided 42 dollars equally among 6 poor families; how many dollars did he give to each?

36. If a drover pay 49 dollars for 7 sheep, how much is that a piece?

37. How many chests of tea could be bought for 50 dollars, at 5 dollars a chest?

38. How many hats, at 6 dollars each, could be bought for 48 dollars?

39. A man bought 56 marbles, which he divided equally among his 7 children; how many did he give to each?

40. How many yards of cloth, at 5 shillings a yard, can I buy for 60 shillings? for 55 shillings?

41. A man having 63 dollars, wished to lay it out in flour, at 7 dollars a barrel; how many barrels could he buy?

42. If I pay 7 dollars a yard for broad-cloth, how many yards can I buy for 63 dollars?

43. If a student read 7 pages a day, how long will it take him to read 70 pages?

44. If a ship sail 6 miles an hour, how long will it take it to sail 54 miles?

45. If a stage-coach run 6 miles an hour, how long will it be in running 60 miles? 66 miles?

46. 7 men built a barn for 77 dollars; how many dollars did 1 man receive, each receiving an equal share?

47. There are 6 working days in 1 week; how many weeks are there in 72 working days?

48. A lady paid 84 shillings for 7 yards of satin; how much was that a yard?

49. A man paid 84 dollars for a quantity of land,

paying at the rate of 7 dollars an acre; how many acres did he buy?

50. How many cows at 7 dollars a piece can I buy for 77 dollars?

51. At 8 dollars a yard, how many yards of cloth can be bought for 16 dollars?

8 in 8,	1	(once)	9 in 9,	1	(once)	10 in 10,	1	(once)
8 " 16,	2	times.	9 " 18,	2	times.	10 " 20,	2	times.
8 " 24,	3	"	9 " 27,	3	"	10 " 30,	3	"
8 " 32,	4	"	9 " 36,	4	"	10 " 40,	4	"
8 " 40,	5	"	9 " 45,	5	"	10 " 50,	5	"
8 " 48,	6	"	9 " 54,	6	"	10 " 60,	6	"
8 " 56,	7	"	9 " 63,	7	"	10 " 70,	7	"
8 " 64,	8	"	9 " 72,	8	"	10 " 80,	8	"
8 " 72,	9	"	9 " 81,	9	"	10 " 90,	9	"
8 " 80,	10	"	9 " 90,	10	"	10 " 100,	10	"
8 " 88,	11	"	9 " 99,	11	"	10 " 110,	11	"
8 " 96,	12	"	9 " 108,	12	"	10 " 120,	12	"

52. A farmer paid 24 dollars for 3 cows; what did they cost him a-piece?

53. If 2 coats cost 18 dollars, what would 1 coat cost?

54. James paid 20 cents for 2 drawing pencils; what would a single one have cost him at the same rate?

55. 8 boys gave equal sums of money to a poor blind man, which amounted in all to 32 cents; how much did each boy give him?

56. A farmer has 3 pastures, each containing the same number of acres; and they altogether contain 27 acres; how many acres in each field?

57. In what length of time could 10 men perform a piece of work which 1 man could do in 30 days?

58. If 9 men can do a piece of work in 5 days, in what time can 3 men do the same work?

59. If 9 barrels of flour are worth 54 dollars, what are 2 barrels worth at the same rate?

60. A farmer has 10 horses, and hay sufficient to feed them for 6 months; how long will the same amount of hay feed 6 horses?

61. How many acres of land at 9 dollars per acre, can be bought for 63 dollars?

62. At 8 dollars per ton, how many tons of hay may be bought for 88 dollars?

63. Allowing a workman to build 9 feet of stone wall in a day, how many days would he require to build 81 feet?

64. At 10 cents a piece, how many plates may be bought for 100 cents?

65. In one dram there are 3 scruples; how many drams in 36 scruples?

66. In one ounce there are 8 drachms; how many ounces in 96 drams?

67. If there are 3 miles in one league, how many leagues are there in 33 miles?

68. How many acres of land at 9 dollars per acre can be bought for 108 dollars? for 81 dollars? 88 dollars?

69. A man having 64 dollars wished to lay it out for sheep, at 8 dollars per head; how many sheep did he buy?

70. If you read 9 pages a day, how many days will you require to read 99 pages? 90? 63?

71. At 10 cents a piece, how many books can you buy for 90 cents? 108? 40? 30? 50?

72. If 9 tons of hay cost 81 dollars, what would one ton cost at the same rate?

73. How many caps at 8 cents a piece can be bought for 56 cents? 64? 72? 80? 88? 96?

74. If 9 pounds of flour will last a family one day, how long will 108 pounds last the same family?

11 in 11,	1 (once.)		12 in 12,	1 (once.)	
11 " 22,	2 times.		12 " 24,	2 times.	
11 " 33,	3 "		12 " 36,	3 "	
11 " 44,	4 "		12 " 48,	4 "	
11 " 55,	5 "		12 " 60,	5 "	
11 " 66,	6 "		12 " 72,	6 "	
11 " 77,	7 "		12 " 84,	7 "	
11 " 88,	8 "		12 " 96,	8 "	
11 " 99,	9 "		12 " 108,	9 "	
11 " 110,	10 "		12 " 120,	10 "	
11 " 121,	11 "		12 " 132,	11 "	
11 " 132,	12 "		12 " 144,	12 "	

75. Bought a piece of cloth containing 11 yards for 22 dollars; how much was it per yard?

76. What is the price of 1 barrel of beef, if 12 barrels cost 24 dollars?

77. If 33 shillings are paid for 11 bushels of apples, what is the cost of one bushel?

78. A father dying, leaves 48 thousand dollars to be equally divided between his twelve children; what will be the share of each child?

79. A certain quantity of provision will last one man 55 days; how long will it last 11 men?

80. When hay is worth 12 dollars a ton, how many tons can be bought for 48 dollars?

81. Eleven men bought a span of horses for 66 dollars; how much did each one pay?

82. A man earns 77 dollars in 11 weeks; how much is that a week?

83. If 12 men can earn 84 dollars in one week, what is each man's share of the money?

84. If you divide 99 dollars equally among 11 men, what will each one receive?

85. A painter received 96 dollars for painting 12 carriages; how much did he receive on each?

86. When coffee is 11 cents a pound, how much can I buy for 110 cents? 121? 132?

87. At 12 shillings a gallon, how many gallons of wine can you buy for 108 shillings?

88. If 1 pound of sugar is worth 12 cents, how many pounds can be bought for 120 cents?

89. A farmer sold 12 cows for 144 dollars; how much did he receive for each?

90. Sixteen is how many times 2? 4? 8?

91. Eighteen is how many times 2? 3? 6? 9?

92. Twenty is how many times 2? 4? 5? 10?

93. Twenty-one is how many times 3? 7?

94. Thirty-six is how many times 2? 3? 4? 6? 9? 12?

95. Forty is how many times 2? 4? 5? 8? 10?

96. Eighty-four is how many times 2? 3? 4? 6? 7? 12?

97. One hundred is how many times 2? 5? 10? 20? 25? 50?

98. Eleven times 8 is how many times 4? 11? 8?

99. Nine times 7 is how many times 9? 3? 7?

100. Six times 6 is how many times 4? 9? 3? 6? 12?

101. Eight times 8 is how many times 2? 4? 9? 7? 5? 11?

Note.—Some few of the foregoing questions may need some simplifying, especially if the pupil has not progressed with a proper degree of thoroughness. The class ought not to be allowed to proceed until the members are perfectly familiar with the foregoing tables, and solve the questions with promptness and precision.

The following section contains questions, some of which embody all of the principles passed over. Be sure they are *understood* by the pupil.

---

§ **16.**—1. If an orange cost 2 cents, and a lemon cost 3 times as much, how many of each an equal number can you buy for 18 cents, and have 2 cents remaining?

2. A man bought 3 sheep at 3 dollars a head, and 2 cows for 12 dollars a head; and after paying 11 dollars in money, gave his note for the remainder, to be paid in wheat at 2 dollars per bushel: what was the amount of the note, and how many bushels of wheat will it take to pay it?

3. A man bought 13 pounds of grass-seed at 2 cents per pound, and 5 yards of cloth at 6 cents a yard; he gave in return 4 pounds of butter, at 12 cents a pound, and the remainder in money; how much money did he pay?

4. A boy had 20 marbles, and gave them to his companions as follows: to one he gave 4, to another 5, and to another 7; his brother then gave him 9, and he then sold all he had for apples—giving one marble for 2 apples; how many apples did he receive?

5. If I buy a sheep for 15 dollars, and pay for it with butter at 3 dollars a firkin, how many firkins will it take?

6. If 7 yards of cloth cost 21 shillings, how many pounds of butter at 2 shillings a pound must I pay for 4 yards of the same cloth?

Note.—First find the price of the cloth per yard. If 7 yards cost 21 shillings, one yard would cost one-seventh of 21 shillings; one-seventh of 21 shillings is 3 shillings: if one yard cost 3 shillings, 4 yards would cost 4 times 3 shillings; 4 times 3 shillings are 12 shillings. It would take as many pounds of butter to pay for it as the number of times 2 shillings, the price of 1 pound, is contained in 12 shillings: 2 shillings in 12 shillings 6 times. Therefore, it would take 6 pounds of butter to pay for the cloth.

7. If 3 pounds of sugar cost 24 cents, what will 5 pounds cost? 6 pounds? 4 pounds? 7 pounds?

8. If 5 barrels of flour cost 20 dollars, what will 3 barrels cost? 4 barrels? 7 barrels? 11 barrels?

9. If 7 oranges are worth 35 cents, what are 5 oranges worth?

10. Two travellers start from the same place and travel in the same direction, one at the rate of 6 miles an hour, and the other 4; how far apart will they be at the end of 3 hours? 4 hours? 7 hours? 12 hours?

11. A. has 4 hours the start of B., and travels at the rate of 5 miles an hour; B. travels 7 miles an hour after him; in what time will B. overtake A.?

12. Seven men bought a quantity of land for 84 dollars, and sold it for 14 dollars less than they gave for it; what did they sell it for, and what was each man's share of the loss?

13. The distance from Madison to Milwaukee is 80 miles; how long will it take a man to travel the distance, going at the rate of ten miles an hour?

---

§ 17. *Note.*—Operations and exercises on numbers are the surest methods of securing mental discipline. As soon as pupils become sufficiently familiar with the principles of Addition, Multiplication, Subtraction, and Division, exercises calculated to hold the mind for a long time on a single question, should be frequently given. In most cases, this will succeed best by having the books of the class closed, and the teacher, or some pupil appointed for the purpose, read the example with a proper degree of slowness. When the exercise is all given, let those who are able to give the answer rise. When all, or nearly all, have arisen, let the teacher or monitor point to some one who shall give the result, when all who agree shall sit. If a variety of answers are given, let the same exercise be gone through with again, in the same manner as before, and repeated until all, or nearly all, arrive at the same result. The teacher should go through the operation with the class, all in silence, until the final result is called for.

1. Four and 3, and 2, and 7, and 5, and 6, and 4, and 11, and 5, and 2, and 8, and 6, and 1, and 10, and 12, and 3, and 5, and 4, and 6, and 3, and 7, and 2, and 8, and 1, and 9, and 2, and 10, and 3, and 11, and 4, and 12, and 5, and 1, and 6, and 2, and 7, and 3, and 8, and 4, and 9, and 5, and 10, and 6, and 11, and 7, and 12, are how many?

2. Five and 3 times 2, and 6, and 4 times 2, and 11, and 3, and 2 less 1, and 5, and 6 divided by 2, and 4, and 7 less 5, and 3 less 1, and 2 times 5, and 9 divided by 3, and 7, and 2, and three times 4 less 6, and 7, and 10, and 8 less 2 times 3, and 9, and 12, and 8, and 10 divided by 5, and 12 less 3 times 4, and 7 less 2, and 3 times 6, less 2 times 5, and 7 less 2, are how many?

3. What is the sum of all the numbers from 1 to 25 inclusive?

4. What is the difference between 12 times 12 and 13 times 13?

Note.—All these examples should be solved *mentally.*

5. There are 12 working hours in a day, 6 working days in a week, and 4 weeks in a month. If a man can earn 2 cents an hour, how many yards of cloth could he buy at 12 cents a yard for a month's wages, after deducting 10 cents a day for his board?

Note.—The questions of this section will be found too difficult for most pupils at the degree of advancement to which they have attained on reaching it, unless great thoroughness has been observed during all the intermediate steps. Their solution requires the closest attention, and a good degree of mental discipline,—not more, however, than every pupil should possess, at this stage of his progress.

CHAPTER V.

§ 18.—Any thing regarded simply as *one* is a UNIT. One book, one dollar, one cent, one apple, &c., are examples.

§ 19.—Units are sometimes broken into parts, and those parts are called FRACTIONS.

If a line should be parted in the center, or an apple cut into two *equal* parts, one of those parts would be called ONE HALF. The two parts would be called TWO HALVES.

Note.—The teacher should illustrate the foregoing by means of visible objects; as an apple, or some other convenient object.

1. If 1 apple is worth 2 cents, what is 1 half of it worth?

Ans. One cent.

2. How many halves in one?

3. How many halves in two?

Ans. Four halves.

4. James divided 3 apples into halves; how many halves were there?

SOLUTION.—In one apple there are 2 halves; in 3 apples there are 3 times 2 halves; 3 times 2 halves are 6 halves. Therefore, in 3 apples there are 6 halves.

5. If you could buy 1 apple for 2 cents, how many could you buy for 3 cents?

Note.—First find how much you could buy for 1 cent.

6. Three is how many times 2?

Ans. Once and 1 half times 2.

7. 4 is how many times 2?

8. 5 is how many times 2?

9. 6 is how many times 2?
10. 7 is how many times 2?
11. 8 is how many times 2?
12. 9 is how many times 2?
13. In 2 apples and one half of an apple, how many halves?
14. If 1 yard of cloth cost 11 cents, what would 1 half of a yard cost? 3 yards? 2 yards and 1 half a yard?
15. If 2 cents will buy 1 apple, how many apples may be bought for 5 cents?

If any number or object be divided into THREE equal parts, one of those parts is called ONE THIRD of the number or object; two parts are called TWO THIRDS; three parts THREE THIRDS, or the whole.

Hence, in 1 there are 3 thirds; in 2 there are 2 times 3 thirds.

16. In 1 and 1 third, how many thirds? in 2? in 2 and 2 thirds? in 3? in 4?
17. How many thirds in 1 apple? in 2 apples? in 3? in 4? in 5? in 6?
18. 6 is how many times 3?
19. 7 is how many times 3?
20. 8 is how many times 3?
21. 9 is how many times 3?
22. 10 is how many times 3?
23. 11 is how many times 3?
24. If a yard of cloth worth 3 dollars, be cut into 3 equal pieces, what will one of the pieces be worth? 2 pieces?
25. What do you understand by 1 third of anything? 2 thirds? 3 thirds?

If any number or object is divided into FOUR equal parts, one of those parts is called ONE FOURTH; two parts, TWO FOURTHS; &c.

26. In 1 yard how many fourths of a yard?

Note.—Fourths are sometimes called QUARTERS.

27. If 1 pound of lard cost 4 cents, what would be the cost of 1 fourth of a pound? 2 fourths or 1 half? 3 fourths?

28. How many fourths in 1 apple and 1 fourth of an apple? 2 apples?

29. How many fourths in 3 apples and 2 fourths of an apple? 4 apples? 5?

30. 5 is how many times 4?

SOLUTION.—As many times 4 as the number of times that 4 is contained in 5: 4 in 5, 1 and 1 fourth times. Therefore, 5 is 1 and 1 fourth times 4.

31. 6 is how many times 4?
32. 7 is how many times 4?
33. 8 is how many times 4?
34. 9 is how many times 4?
35. 10 is how many times 4?
36. 11 is how many times 4?

37. If 12 pounds of coffee be divided equally among 4 persons, how many pounds will each receive?

38. Three is what part of 12? of 6? of 9?

If any number or object is divided into FIVE equal parts, one of those parts is called ONE FIFTH; two parts TWO FIFTHS; &c.

39. If a piece of cloth worth 5 dollars be equally divided among 5 men, what will each one's share be worth?

40. What is 1 fifth of 5? of 6? of 7? of 8? of 9? of 10? of 11? of 12?

41. What do you understand by 1 third, 1 fourth, 1 fifth, &c., of anything?

When a number is divided into 6 equal parts, one of those parts is called 1 sixth; when into 7, sevenths; 8, eighths; 9, ninths; 10, tenths; &c.

42. When wheat is worth 6 shillings a bushel, what

part of a bushel can you buy for 1 shilling? for 2 shillings? for 3? for 4? for 5? for 6?

43. At 7 dollars a barrel, what part of a barrel of flour could you buy for 1 dollar? for 2 dollars? for 3? for 4? for 5? for 6? for 7?

44. If 8 men can do a job of work in 1 day, what part of it could 1 man do? 2 men? 3 men? 4? 5? 6? 7? 8?

45. If a man can travel 18 miles in 9 hours, how far could he travel in 1 hour? in 2 hours? in 5 hours? in 8 hours?

46. What do you understand by 1 ninth of anything? 1 tenth? &c.

47. 12 is how many times 6?

48. 13 is how many times 6?

49. 15 is how many times 6?

50. 17 is how many times 6?

51. 18 is how many times 6?

52. 19 is how many times 6?

53. 22 is how many times 6?

54. 24 is how many times 6?

55. 26 is how many times 6?

56. 14 is how many times 7?

57. 17 is how many times 7?

58. 16 is how many times 7?

59. 25 is how many times 7?

60. 19 is how many times 8?

61. 24 is how many times 8?

62. 27 is how many times 8?

63. 30 is how many times 8?

64. 32 is how many times 8?

65. How much is 1 tenth of 10? of 20? of 40? of 50?

Note.—The question, what do you understand by 1 tenth, &c., of anything, should be often repeated, and great care taken to see that the terms are clearly comprehended by the pupil.

66. What is 1 half of 12? How do you find 1 half of any number?

67. What is 1 fourth of 13? 14? 15? 16? 17? 20? 25?

68. What is 1 seventh of 8? 11? 14? 19? 21? 30? 40?

69. What is 1 ninth of 18? 22? 26? 17? 11? 40? 55?

70. What is 1 third of 16? 21? 26? 18? 13? 32? 20?

71. What is 1 fifth of 12? 16? 20? 11? 19? 50? 27?

72. What is 1 seventh of 11? 14? 36? 27? 37? 61? 70?

73. What is 1 eighth of 8? 16? 26? 28? 39? 72? 64?

74. What is 1 sixth of 17? 19? 30? 41? 52? 63? 75?

75. What is 1 tenth of 18? 21? 36? 42? 57? 80? 100?

76. For 63 dollars, how many barrels of flour can be bought at 7 dollars a barrel? at 9 dollars? at 6 dollars? at 8 dollars?

77. 12 times 1 is how many times 2? 3? 5?

78. 17 times 1 is how many times 3? 4? 7? 8? 10?

79. 18 is how many times 3? 5? 7? 9? 8? 6? 4? 2?

80. 13 is how many times 10? 4? 3? 5? 7? 8? 6? 9? 11?

81. 15 is how many times 5? 6? 3? 7? 2? 9? 8? 11? 10? 12?

82. 14 is how many times 2? 7? 4? 9? 6? 5? 8? 11? 10? 12?

83. 16 is how many times 3? 5? 7? 2? 4? 6? 9? 8? 11? 10?

84. 19 is how many times 4? 6? 5? 3? 8? 7? 10? 11? 9? 12?

85. 12 is how many times 6? 3? 8? 2? 9? 11? 12? 11? 3? 5?

86. 21 is how many times 7? 4? 9? 3? 10? 5? 11? 2? 12? 6?

87. 30 is how many times 8? 5? 2? 4? 3? 6? 7? 9? 11? 12?

88. 22 is how many times 9? 2? 8? 3? 6? 10? 11? 7? 4? 12? 5?

89. 29 is how many times 3? 5? 4? 6? 8? 7? 10? 11? 12? 10? 2?

90. 23 is how many times 4? 2? 5? 7? 3? 6? 8? 11? 9? 10? 12?

91. 28 is how many times 5? 3? 2? 4? 6? 8? 7? 10? 11? 9? 12?

92. 24 is how many times 6? 2? 3? 5? 6? 4? 8? 10? 7? 11? 12?

93. 27 is how many times 7? 3? 2? 6? 4? 8? 5? 10? 11? 12? 9?

94. 25 is how many times 8? 2? 4? 3? 9? 6? 8? 11? 10? 12? 5?

95. 31 is how many times 9? 3? 2? 4? 6? 7? 5? 11? 12? 8? 10?

96. 26 is how many times 2? 4? 6? 8? 3? 5? 7? 12? 10? 11? 9?

97. 37 is how many times 3? 2? 8? 9? 4? 10? 5? 6? 11? 12? 7?

98. 40 is how many times 7? 9? 2? 6? 10? 12? 8? 4? 11? 3? 5?

99. 33 is how many times 12? 3? 5? 8? 7? 6? 9? 11? 10? 2? 3?

100. 39 is how many times 10? 2? 4? 3? 6? 11? 9? 12? 8? 7? 5?

101. 36 is how many times 5? 10? 6? 3? 8? 11? 12? 7? 9? 2? 4?

102. 38 is how many times 4? 2? 5? 7? 3? 6? 8? 10? 11? 9? 12?

103. 37 is how many times 12? 11? 9? 10? 8? 6? 7? 2? 4? 5? 3?

104. 41 is how many times 10? 12? 11? 9? 6? 8? 5? 4? 3? 2? 7?

105. 50 are how many times 5? 10? 7? 3? 12? 4? 6?

106. A drover has 60 dollars to lay out in sheep; how many can he buy, at 2 dollars a head? How many at 3 dollars a head? 10 dollars a head? 5 dollars? 12 dollars?

107. At 4 dollars a barrel, how many barrels of flour can be bought for 42 dollars? How many barrels at 12 dollars? at 5 dollars? at 7 dollars? at 8 dollars? at 6 dollars? at 11 dollars? at 9 dollars? at 10 dollars? at 2 dollars?

108. 44 is how many times 12? 9? 11? 6? 4? 3? 5? 7? 8?

Solution.—44 is as many times 12 as the number of times that 12 is contained in 44; 12 in 44, 3 and 8 twelfth times. Therefore, 44 is 3 and 8 twelfth times 12. 4 and 8 ninth times 9. 4 times 11. 7 and 2 sixth times 6. 11 times 4. 14 and 2 third times 3. 8 and 4 fifth times 5. 6 and 2 seventh times 7. 5 and 4 eighth times 8.

Note.—It would be well to require the full solution for each number, as given for how many times 12, until the pupils are perfectly familiar with the form; after which, simply the answer will be sufficient. It is best, however, to insist on the full solution in the answer to the first question in each example.

109. 55 is how many times 4? 8? 10? 11? 3? 7? 5? 6? 9?

110. 43 is how many times 3? 9? 11? 12? 4? 8? 7? 5? 6?

111. 58 is how many times 5? 10? 12? 3? 6? 7? 8? 4? 9?

112. 52 is how many times 8? 11? 3? 4? 7? 8? 9? 5? 6?

113. 47 is how many times 7? 12? 4? 8? 5? 9? 10? 6? 3?

114. 53 is how many times 6? 3? 5? 9? 6? 10? 11? 7? 4?

115. 45 is how many times 9? 4? 6? 10? 7? 11? 12? 8? 5?

116. 54 is how many times 10? 5? 7? 11? 8? 12? 3? 9? 6?

117. 46 is how many times 8? 6? 9? 12? 10? 3? 4? 7? 11?

118. 56 is how many times 11? 9? 6? 4? 7? 5? 12? 8? 3?

119. 57 is how many times 11? 10? 9? 5? 6? 3? 8? 4? 12?

120. 48 is how many times 12? 11? 8? 6? 7? 3? 5? 10? 4?

121. 58 is how many times 11? 5? 9? 7? 8? 10? 6? 12? 3?

122. 49 is how many times 6? 4? 6? 5? 8? 7? 12? 9? 11?

123. 59 is how many times 8? 3? 5? 4? 9? 7? 10? 11? 12?

124. 61 is how many times 7? 3? 6? 12? 9? 10? 4? 11? 5?

125. 69 is how many times 3? 4? 6? 12? 11? 10? 9? 8? 7?

126. 60 is how many times 10? 7? 12? 9? 3? 5? 4? 6? 11?

127. 67 is how many times 8? 4? 12? 9? 11? 10? 8? 7? 6?

128. 63 is how many times 9? 12? 10? 9? 8? 7? 6? 3? 5?

129. 68 is how many times 5? 11? 12? 8? 9? 6? 7? 10? 3?

130. 62 is how many times 5? 12? 6? 9? 11? 3? 8? 10? 4?

131. 65 is how many times 6? 4? 11? 10? 8? 7? 3? 6? 9?

132. 64 is how many times 5? 12? 10? 9? 7? 8? 3? 4? 6?

133. How many pounds of sugar, at 8 cents a pound, can you buy for 25 cents? for 35 cents? for 45 cents? for 55 cents?

134. In 65 shillings, how many dollars, New-York currency?

135. A drover had 96 dollars, which he paid for sheep at 7 dollars a head; how many sheep did he buy? How many could he have bought at 9 dollars a head?

136. A man hired a horse and agreed to give 5 cents a mile for his use; he paid 56 cents; how many miles did he ride?

137. A traveler can perform a journey in 45 hours; how many days will it take him to do it when the days are 11 hours long?

138. A hare has 87 rods the start of a grey-hound, but every 2 minutes the hound runs 19 rods while the hare runs 12; how many minutes will it require the hound to overtake the hare?

139. A cask containing 67 gallons, has a tube leading from it that will discharge 7 gallons every 5 minutes; in how many minutes will the cask be emptied?

140. In a certain school containing 98 scholars, one eleventh are learning geography; 1 sixth are learning arithmetic, and the remainder are learning to read; how many are there attending to each study?

141. Five men bought a horse for 52 dollars; what was each man's share? They afterwards sold the

horse for 73 dollars; what was each man's share of the gain?

142. A farmer put 85 pounds of butter into 11 boxes; how many pounds did he put into each box?

143. If a man can do a piece of work in 98 hours, how many days of 12 hours each, will it require for him to do it?

CHAPTER VI.

§ 20.—We have already learned, (§ 19,) that parts of a unit (§ 18) are called fractions. Fractions, as well as whole numbers, may be represented by figures. To do this, two numbers are required; one to show the number or size of the parts into which the unit is broken, and the other to show how many of those parts are taken. These numbers are usually written one above the other with a line between them. The number below the line shows into how many parts the unit is divided or the size of the parts, and is called the DENOMINATOR. The number above the line shows how many parts are taken, and is called the NUMERATOR.

ONE HALF is represented by the character $\frac{1}{2}$.

One third is written $\frac{1}{3}$; two thirds $\frac{2}{3}$; three thirds $\frac{3}{3}$, which equals 1.

One fourth is written $\frac{1}{4}$; two fourths $\frac{2}{4}$; three fourths $\frac{3}{4}$.

One fifth is written $\frac{1}{5}$; two fifths $\frac{2}{5}$; three fifths $\frac{3}{5}$, &c., &c.

One sixth is written $\frac{1}{6}$; two sixths $\frac{2}{6}$, &c., &c.

Illustration.—$\frac{5}{7}$ of an apple denotes that the apple is cut into 7 pieces, and that 5 of those parts are taken.

Note.—The teacher should illustrate by actually cutting an apple or some other convenient object into seven equal parts.

§ **21.**—It will be observed that the fraction takes its name from the *denominator;* (§ 20;) thus when the unit is divided into *eight* equal parts, they are called *eighths*, and one part is represented $\frac{1}{8}$; when the unit is divided into *nine* equal parts, the parts are called *ninths;* and one part is represented $\frac{1}{9}$, &c., &c.

Note.—Care must be taken to make the pupils familiar with the mode of expressing fractions by figures. Be sure they understand fully what is denoted by the denominator, which should be first explained; also by the numerator. The pupil should be required to write the fractions on the black-board, and made frequently to explain the meaning of each term.

§ **22.**—If the *numerator* is *larger* than the *denominator*, the value of the fraction is greater than a unit, and the fraction is called IMPROPER. If the *numerator* is *less* than the *denominator*, the value of the fraction is *less* than a unit, and the fraction is called PROPER.

REVIEW.—What are fractions? How are they expressed by figures? What is the number below the line called and what does it denote?—the number above the line? From which of these numbers does the fraction take its name? Why? What is an improper fraction? What is a proper fraction? Will you write the fraction one ninth on the black-board? One seventh? One fourth? Five sevenths? Eight tenths? Nine fifths? Seven elevenths? Five thirds, &c., &c.

Is nine tenths a proper, or an improper fraction? Why? Is $\frac{3}{4}$ a proper or an improper fraction? $\frac{4}{2}$? $\frac{3}{5}$? $\frac{7}{9}$? $\frac{8}{6}$? $\frac{11}{9}$? $\frac{11}{12}$? &c.

---

1. If $\frac{1}{2}$ of an orange cost 3 cents, how much will a whole orange cost?

SOLUTION.—If $\frac{1}{2}$ of an orange cost 3 cents, two, the number of halves in 1, will cost 2 times 3 cents; 2 times 3 cents are 6 cents. Therefore, one orange would cost 6 cents.

2. A man sold $\frac{1}{3}$ of a cord of wood for 4 shillings; how much would a whole cord cost? In one cord of wood, how many thirds of a cord? how many fourths? how many sevenths? ninths? halves?

3. At 4 dollars a barrel, what will 3 and $\frac{1}{4}$ barrels of flour cost? 3 times 4 and $\frac{1}{4}$ of 4 are how many?

4. If a laborer can earn 5 dollars a week, how much can he earn in 4 weeks and $\frac{2}{5}$ of a week? 4 times 5 and $\frac{2}{5}$ of 5 are how many?

5. 3 times 6 and $\frac{5}{6}$ of 6 are how many?

6. 4 times 7 and $\frac{3}{7}$ of 7 are how many?

7. 5 times 8 and $\frac{2}{8}$ of 8 are how many?

8. 6 times 9 and $\frac{1}{9}$ of 9 are how many?

9. 7 times 10 and $\frac{3}{10}$ of 10 are how many?

10. At 12 cents a pound, what would 8 pounds and $\frac{1}{2}$ a pound of sugar cost?

11. 9 times 11 and $\frac{3}{11}$ of 11 are how many?

12. 10 times 12 and $\frac{2}{12}$ of 12 are how many?

13. 4 times 9 and $\frac{2}{9}$ of 9 are how many?

14. 7 times 8 and $\frac{3}{8}$ of 8 are how many?

15. 8 times 5 and $\frac{4}{5}$ of 5 are how many?

16. A man bought 3 barrels of cider at 4 dollars a barrel, and to pay for it he gave flour at 3 dollars a barrel; how many barrels did it take? 3 times 4 is how many times 3?

17. A lady bought 4 yards of cloth, at $2\frac{1}{2}$ dollars a yard, and paid for it with thread at 2 dollars a pound; how many pounds of thread did it take? 4 times $2\frac{1}{2}$ is how many times 2?

18. 5 times 6 and $\frac{1}{6}$ of 6 are how many times 4? 7? 3? 9?

19. 7 times 7 and $\frac{2}{7}$ of 7 are how many times 5? 6? 8? 9? 10? 11? 12?

20. 9 times 10 and $\frac{3}{10}$ of 10 are how many times 12? 7? 8? 11?

21. 3 times 12 and $\frac{7}{12}$ of 12 are how many times 7? 6? 9? 10? 5? 4?

22. 8 times 5 and $\frac{4}{5}$ of 5 are how many times 6? 9? 3? 8? 7? 5?

SOLUTION.—8 times 5 is 40. $\frac{1}{5}$ of 5 is 1; $\frac{4}{5}$ is 4 times 1; 4 times 1 is 4; 40 and 4 are 44. 44 is as many times 6 as the number of times that 6 is contained in 44; 6 in 44, $7\frac{2}{6}$ times. Therefore, 8 times 5 and $\frac{4}{5}$ of 5 are $7\frac{2}{6}$ times 6.

23. 9 times 6 and $\frac{1}{6}$ of six are how many times 7? 4? 8?

24. 3 times 4 and $\frac{1}{4}$ of 4 are how many times 2? 5? 6? 7?

25. 10 times 5 and $\frac{3}{5}$ of 5 are how many times 7? 8? 9? 11? 12?

26. 8 times 9 and $\frac{2}{9}$ of 9 are how many times 5? 6? 7? 8? 10?

27. 5 times 10 and $\frac{5}{10}$ of 10 are how many times 7? 6? 8? 7? 12?

28. 11 times 11 and $\frac{1}{11}$ of 11 are how many times 7? 8? 6? 9? 10?

29. 12 times 12 and $\frac{1}{12}$ of 12 are how many times 6? 7? 8? 9? 11?

30. 5 times 12 and $\frac{5}{12}$ of 12 are how many times 7? 6? 9? 8? 11?

---

31. If 3 yards of cloth cost 9 dollars, what will 4 yards cost?

32. If 3 fourths of a barrel of flour cost 6 dollars, what is that a barrel?

33. If I pay 12 dollars for 6 sheep, what must I pay for 7, at the same rate? 8? 9? 12? 11?

34. If 3 men can do a piece of work in 5 days, in what length of time can 7 men do it? 4 men? 10 men?

35. If 27 dollars are paid for 9 barrels of cider, how many yards of cloth, at 2 dollars a yard, must be given for 7 barrels of cider?

---

§ **23.**—1. How many fourths are there in 1? 2? $3\frac{1}{4}$? $5\frac{3}{4}$? $7\frac{2}{4}$? $9\frac{1}{4}$?

2. How many fifths in 1? in $2\frac{1}{5}$? $3\frac{3}{5}$? $4\frac{1}{5}$? $5\frac{4}{5}$? $7\frac{2}{5}$? $8\frac{4}{5}$? $10\frac{1}{5}$?

3. How many sixths in $3\frac{1}{6}$? 2? $5\frac{5}{6}$? $7\frac{4}{6}$? $10\frac{1}{6}$? $8\frac{3}{6}$? $7\frac{7}{6}$?

4. How many sevenths in 3? $4\frac{1}{7}$? $3\frac{1}{7}$? $8\frac{3}{7}$? $11\frac{5}{7}$? $12\frac{6}{7}$? $9\frac{5}{7}$?

5. How many eighths in $2\frac{1}{8}$? $5\frac{1}{8}$? $7\frac{3}{8}$? $8\frac{3}{8}$? $9\frac{1}{8}$? $7\frac{7}{8}$? $3\frac{1}{8}$?

---

§ **24.**—1. How many whole ones in $\frac{2}{2}$? in $\frac{3}{2}$? $\frac{7}{2}$? $\frac{12}{2}$? $\frac{25}{2}$?

2. How many whole ones in $\frac{3}{3}$? $\frac{9}{3}$? $\frac{11}{3}$? $\frac{27}{3}$? $\frac{31}{3}$? $\frac{47}{3}$? $\frac{50}{3}$?

3. How many whole ones in $\frac{6}{6}$? $\frac{21}{6}$? $\frac{37}{6}$? $\frac{42}{6}$? $\frac{54}{6}$? $\frac{61}{6}$? $\frac{45}{6}$?

4. How many whole ones in $\frac{9}{9}$? $\frac{20}{9}$? $\frac{40}{9}$? $\frac{51}{9}$? $\frac{37}{9}$? $\frac{97}{9}$? $\frac{86}{9}$?

5. How many whole ones in $\frac{10}{10}$? $\frac{13}{10}$? $\frac{55}{10}$? $\frac{84}{10}$? $\frac{96}{10}$? $\frac{99}{10}$?

---

§ **25.**—1. James gave apples to his companions as follows: to one he gave $3\frac{1}{4}$; to another $5\frac{2}{4}$; to another $2\frac{1}{4}$; and to another $6\frac{3}{4}$; how many did he give away?

2. A farmer had two pastures; one containing $13\frac{5}{11}$ acres, and the other $11\frac{7}{11}$ acres; how many acres in both?

3. A man bought $7\frac{3}{12}$ barrels of cider and sold $5\frac{1}{12}$ barrels; how much had he remaining?

4. In one rod there are $16\frac{1}{2}$ feet, and 6 feet in one fathom; how many more feet in a rod than in a fathom?

5. What is the sum and difference of 3 rods and 2 rods, in feet?

---

§ **26.**—1. How many barrels of flour, at 3 dollars a barrel, must be given for 5 boxes of butter, at 4 dollars a box?

2. When wood is 6 dollars a cord, how many cords can be bought with 13 yards of cloth at 2 dollars a yard?

3. Bought 11 yards of canvas for 5 cents a yard; how many dimes did it come to?

4. How many barrels of salt, at 3 dollars a barrel, will be required to pay for $7\frac{3}{5}$ pounds of opium at 5 dollars a pound?

5. How much wheat, at 6 shillings a bushel, can be bought for 7 barrels of cider at 3 dollars a barrel?

6. If a laborer can earn 5 dollars in 3 days, how much can he earn in 3 weeks, there being 6 working days in a week?

7. How much must you pay for 8 pounds of beef, at $5\frac{1}{3}$ cents a pound? at $3\frac{1}{2}$ cents? $2\frac{2}{3}$ cents?

8. How many pounds of coffee, at 8 cents a pound, must be given for $9\frac{1}{2}$ pounds of sugar at 10 cents a pound?

9. A boy bought $10\frac{3}{7}$ pounds of butter at 6 cents a pound, and paid for it with lard at 5 cents a pound; how many pounds did it take?

10. How many figs, at 2 for 3 cents, can you buy for 27 cents? 18 cents? 13 cents? 19 cents?

---

§ **27.**—1. If $\frac{1}{2}$ of an apple cost 1 cent, what will 1 apple cost? 2 apples? 3 apples?

2. If $\frac{2}{3}$ of a yard of cloth cost 4 dollars, what will be the cost of 2 yards? 3 yards? 4 yards?

3. 4 is $\frac{2}{3}$ of what number?

4. If $\frac{3}{4}$ of a barrel of flour cost 6 dollars, what will 3 barrels cost?

SOLUTION.—If $\frac{3}{4}$ of a barrel cost 6 dollars, $\frac{1}{4}$ will cost $\frac{1}{3}$ of 6 dollars; $\frac{1}{3}$ of 6 dollars is 2 dollars; 4, the number of fourths in 1, will cost 4 times 2 dollars; 4 times 2 dollars are 8 dollars; 3 barrels will cost 3 times 8 dollars; 3 times 8 dollars are 24 dollars:—Therefore, if $\frac{3}{4}$ of a barrel of flour cost 6 dollars, 3 barrels will cost 24 dollars.

5. If $\frac{2}{5}$ of a hundred weight of aquafortis cost 10 dollars, what would 2 pounds cost? 10 is $\frac{2}{5}$ of what number?

6. Bought $\frac{1}{7}$ of a ton of hay for 2 dollars; what would 3 tons cost at the same rate? 2 is $\frac{1}{7}$ of what number?

7. 2 is $\frac{1}{7}$ of what number?
8. 9 is $\frac{1}{2}$ of what number?
9. 5 is $\frac{1}{3}$ of what number?
10. 7 is $\frac{1}{9}$ of what number?
11. 4 is $\frac{1}{4}$ of what number?
12. 6 is $\frac{1}{8}$ of what number?
13. 3 is $\frac{1}{11}$ of what number?
14. 8 is $\frac{1}{5}$ of what number?

15. 3 is $\frac{1}{9}$ of what number?
16. 4 is $\frac{1}{12}$ of what number?
17. 9 is $\frac{3}{4}$ of what number?
18. 8 is $\frac{2}{3}$ of what number?
19. 10 is $\frac{2}{5}$ of what number?
20. 6 is $\frac{2}{7}$ of what number?
21. 12 is $\frac{4}{11}$ of what number?
22. 10 is $\frac{2}{7}$ of what number?
23. 5 is $\frac{5}{7}$ of what number?
24. 8 is $\frac{4}{10}$ of what number?
25. 12 is $\frac{6}{8}$ of what number?
26. 14 is $\frac{2}{7}$ of what number?
27. 16 is $\frac{8}{9}$ of what number?
28. 15 is $\frac{3}{5}$ of what number?
29. 18 is $\frac{6}{11}$ of what number?
30. 18 is $\frac{3}{10}$ of what number?
31. 27 is $\frac{9}{12}$ of what number?
32. 25 is $\frac{5}{9}$ of what number?
33. 33 is $\frac{11}{5}$ of what number?
34. 40 is $\frac{8}{7}$ of what number?
35. 50 is $\frac{10}{9}$ of what number?
36. 36 is $\frac{12}{5}$ of what number?
37. 42 is $\frac{7}{2}$ of what number?
38. 28 is $\frac{4}{12}$ of what number?
39. 30 is $\frac{3}{8}$ of what number?
40. 44 is $\frac{3}{11}$ of what number?
41. 64 is $\frac{12}{11}$ of what number?
42. 63 is $\frac{7}{9}$ of what number?
43. 84 is $\frac{7}{12}$ of what number?
44. 96 is $\frac{12}{9}$ of what number?
45. 144 is $\frac{12}{5}$ of what number?

SOLUTION.—If 144 is $\frac{12}{5}$, $\frac{1}{12}$ of 144 is $\frac{1}{5}$; $\frac{1}{12}$ of 144 is 12; if 12 is $\frac{1}{5}$, 5 times 12 is 5 fifths; 5 times 12 is 60:—Therefore, 144 is $\frac{12}{5}$ of 60.

46. 60 is $\frac{5}{12}$ of what number?

SOLUTION.—If 60 is $\frac{5}{12}$, $\frac{1}{5}$ of 60 is $\frac{1}{12}$; $\frac{1}{5}$ of 60 is 12; if 12 is $\frac{1}{12}$, 12 times 12 is 12 twelfths; 12 times 12 is 144 :—Therefore, 60 is $\frac{5}{12}$ of 144.

Note.—It will be observed that the 46th question is the reverse of the 45th, and that they reciprocally prove each other. It will be found a valuable exercise to have the pupils in the same manner reverse each operation. It should be insisted on, until the mode of solution and the principle involved is rendered perfectly familiar.

47. 24 is $\frac{8}{9}$ of what number?
48. 49 is $\frac{7}{8}$ of what number?
49. 20 is $\frac{5}{7}$ of what number?
50. 56 is $\frac{7}{8}$ of what number?

---

51. A man sold a cow for 12 dollars, which was $\frac{3}{4}$ of what she cost him; he paid for her with cloth at $\frac{1}{3}$ of 9 dollars a yard; how much did the cow cost him, and how many yards of cloth did it take to pay for her?

52. A man bought a horse, and being asked the price of it, answered, that he offered 36 dollars, which was $\frac{6}{7}$ of what the man asked him for it, and that he gave 2 dollars less than the price asked; what did he pay for the horse?

53. If $\frac{6}{7}$ of a cask of wine cost 36 dollars, what will be the cost of 2 casks? How many barrels of flour, at 5 dollars a barrel, would it take to pay for it?

54. 9 is $\frac{3}{8}$ of some number; what is $\frac{1}{6}$ of the same number?

55. 10 is $\frac{5}{8}$ of some number; what is $\frac{1}{2}$ of the same number?

SOLUTION.—If 10 is $\frac{5}{8}$, $\frac{1}{5}$ of 10 is $\frac{1}{8}$; $\frac{1}{5}$ of 10 is 2; if 2 is $\frac{1}{8}$, 8 times 2 is 8 eighths; 8 times 2 is 16. $\frac{1}{2}$ of 16 is 8. Therefore, 8 is $\frac{1}{2}$ of that number of which 10 is $\frac{5}{8}$.

56. If 20 is $\frac{5}{8}$ of some number, what is $\frac{2}{4}$ of the same number?

57. There is a pole standing in the water, $\frac{1}{5}$ of it is in the mud, and 2 feet, which is $\frac{1}{5}$ of the whole pole, is above the water; what is the whole length of the pole?

58. There is a school in which $\frac{3}{11}$ learn grammar, $\frac{1}{11}$ learn algebra, $\frac{2}{11}$ learn to read, $\frac{4}{11}$ learn arithmetic, and 5 learn to write; how many scholars are in the school, and how many attending to each branch?

59. A drover sold a horse for 84 dollars, which was $\frac{8}{5}$ of what it cost him; how much did he gain by the bargain, and how much did the horse cost him?

Note.—Questions of this nature may be solved by first finding one fifth, and then the cost of the horse, which subtract from the price received, to get the gain; or a preferable mode is to say he gained $\frac{3}{5}$, then, having the value of $\frac{1}{5}$, the gain is easily found. Either form may be used.

60. 25 is $\frac{5}{7}$ of how many times 9? 5? 7?
61. 27 is $\frac{3}{7}$ of how many times 5? 4? 3?
62. 30 is $\frac{3}{4}$ of how many times 8? 6? 5?
63. 36 is $\frac{9}{8}$ of how many times 5? 7? 10?
64. 42 is $\frac{6}{7}$ of how many times 4? 6? 8?
65. 45 is $\frac{9}{10}$ of how many times 6? 9? 11?
66. 56 is $\frac{7}{9}$ of how many times 10? 8? 5?
67. 64 is $\frac{8}{11}$ of how many times 6? 4? 9?
68. 63 is $\frac{9}{11}$ of how many times 7? 10? 12?
69. 70 is $\frac{10}{4}$ of how many times 3? 8? 7?
70. 72 is $\frac{12}{7}$ of how many times 6? 5? 11?
71. 80 is $\frac{10}{4}$ of how many times 3? 9? 5?
72. 84 is $\frac{7}{2}$ of how many times 8? 6? 3?
73. 96 is $\frac{8}{3}$ of how many times 9? 4? 7?
74. 108 is $\frac{9}{5}$ of how many times 8? 11? 7?
75. 144 is $\frac{12}{7}$ of how many times 5? 4? 3?

Note.—In solving the foregoing questions, the pupil should not be permitted to ask the question the second time. After the solution is commenced, it should not be interrupted.

---

76. If $\frac{2}{3}$ of an orange cost 4 cents, what will $\frac{1}{3}$ cost?

SOLUTION.—If $\frac{2}{3}$ cost 4 cents, $\frac{1}{3}$ will cost $\frac{1}{2}$ of 4 cents; $\frac{1}{2}$ of 4 cents is 2 cents. Therefore, if $\frac{2}{3}$ of an orange cost 4 cents, $\frac{1}{3}$ will cost 2 cents.

77. 6 is $\frac{2}{4}$ of some number; what is $\frac{1}{4}$ of the same number?

78. 20 is $\frac{5}{7}$ of some number; what is $\frac{2}{7}$ of the same number?

79. 24 is 3 times what number? 4 times what number?

80. If a man can earn 60 cents a day by working $\frac{2}{3}$ of the time, how much could he earn by working all the time? 60 is $\frac{2}{3}$ of what number?

81. There is a pole $\frac{2}{7}$ in the mud, $\frac{3}{7}$ in the water, and 6 feet out of the water; how long is the pole? How much in the water, and how much in the mud?

82. A farmer sold a horse for 63 dollars, which was $\frac{7}{5}$ of what it cost him; how much did he gain by the bargain?

SOLUTION.—He gained $\frac{2}{5}$. If 63 dollars are $\frac{7}{5}$, $\frac{1}{7}$ of 63 dollars is $\frac{1}{5}$; $\frac{1}{7}$ of 63 dollars is 9 dollars; if 9 dollars is $\frac{1}{5}$, 2 times 9 dollars are $\frac{2}{5}$; 2 times 9 dollars are 18 dollars. Therefore, he gained 18 dollars.

83. A man sold a village lot for 84 dollars, by which he lost $\frac{2}{7}$ of what it cost him; how much did it cost, and how much did he lose by the sale?

84. A man sold a cow for 27 dollars, and gained $\frac{1}{8}$ by the sale; what did the cow cost him?

85. A lady lost $\frac{3}{5}$ of her money, and then had 27 dollars remaining in her purse; how much did she lose?

CHAPTER VII.

§ 28.—1. A man being asked the age of his wife, answered, that his eldest daughter was 6 years old, and that $\frac{2}{3}$ of her age was just $\frac{1}{7}$ of the age of his wife; what was her age? $\frac{2}{3}$ of 6 is $\frac{1}{7}$ of what number?

Solution.—$\frac{1}{3}$ of 6 is 2; $\frac{2}{3}$ is 2 times 2; 2 times 2 is 4; if 4 is $\frac{1}{7}$, 7 times 4 is $\frac{7}{7}$; 7 times 4 is 28. Therefore, $\frac{2}{3}$ of 6 is $\frac{1}{7}$ of 28.

2. Two men, talking of their sheep, one said he had 18. Then, said the other, $\frac{3}{7}$ of mine is exactly $\frac{5}{6}$ of yours; how many had he? $\frac{5}{6}$ of 18 is $\frac{3}{7}$ of what number?

3. A drover bought a sheep for 16 dollars, and sold it again for $\frac{6}{7}$ of $\frac{3}{4}$ of the first cost; did he gain or lose by the bargain, and how much? $\frac{3}{4}$ of 16 is $\frac{6}{7}$ of what number?

4. $\frac{4}{5}$ of 20 is $\frac{2}{3}$ of what number?
5. $\frac{5}{7}$ of 28 is $\frac{4}{5}$ of what number?
6. $\frac{9}{10}$ of 40 is $\frac{6}{7}$ of what number?
7. $\frac{4}{9}$ of 45 is $\frac{5}{11}$ of what number?
8. $\frac{8}{11}$ of 33 is $\frac{4}{5}$ of what number?
9. $\frac{5}{12}$ of 72 is $\frac{6}{9}$ of what number?
10. $\frac{7}{10}$ of 80 is $\frac{8}{11}$ of what number?
11. $\frac{3}{8}$ of 32 is $\frac{2}{9}$ of how many times 6?

Solution.—$\frac{1}{8}$ of 32 is 4; $\frac{3}{8}$ is 3 times 4; 3 times 4 is 12; if 12 is $\frac{2}{9}$, $\frac{1}{2}$ of 12 is $\frac{1}{9}$; $\frac{1}{2}$ of 12 is 6; if 6 is $\frac{1}{9}$, nine times 6 is $\frac{9}{9}$; 9 times 6 is 54; as many times 6 as the number of times that 6 is contained in 54; 6 in 54, 9 times. Therefore, $\frac{3}{8}$ of 32 is $\frac{2}{9}$ of 9 times 6.

12. $\frac{6}{8}$ of 32 is $\frac{8}{9}$ of how many times 4? 5? 3?

13. $\frac{3}{5}$ of 35 is $\frac{7}{8}$ of how many times 6? 7? 9?
14. $\frac{5}{8}$ of 40 is $\frac{5}{9}$ of how many times 7? 9? 11?
15. $\frac{4}{6}$ of 48 is $\frac{2}{5}$ of how many times 5? 3? 7?
16. $\frac{7}{8}$ of 56 is $\frac{7}{9}$ of how many times 3? 5? 8?
17. $\frac{4}{7}$ of 63 is $\frac{6}{5}$ of how many times 6? 7? 8?
18. $\frac{5}{9}$ of 72 is $\frac{3}{7}$ of how many times 5? 9? 7?
19. $\frac{8}{9}$ of 81 is $\frac{6}{10}$ of how many times 7? 5? 8?
20. $\frac{6}{8}$ of 96 is $\frac{9}{11}$ of how many times 8? 7? 10?

---

§ 29.—1. 2 thirds of 9 is $\frac{3}{7}$ of how many fifths of 20?

Solution.—$\frac{1}{3}$ of 9 is 3; $\frac{2}{3}$ is 2 times 3; 2 times 3 is 6; if 6 is $\frac{3}{7}$, $\frac{1}{3}$ of 6 is $\frac{1}{7}$; $\frac{1}{3}$ of 6 is 2; if 2 is $\frac{1}{7}$, 7 times 2 is $\frac{7}{7}$; 7 times 2 is 14; as many fifths of 20 as the number of times that $\frac{1}{5}$ of 20 is contained in 14; $\frac{1}{5}$ of 20 is 4; 4 in 14, $3\frac{2}{4}$ times. Therefore, $\frac{2}{3}$ of 9 is $\frac{3}{7}$ of $3\frac{2}{4}$ fifths of 20.

Note.—3 and $\frac{2}{4}$ fifths, is called a *complex* fraction, and is written $\frac{3\frac{2}{4}}{5}$.

2. $\frac{4}{5}$ of 20 is $\frac{8}{9}$ of how many sevenths of 35?
3. $\frac{6}{7}$ of 21 is $\frac{2}{3}$ of how many sixths of 24?
4. $\frac{7}{5}$ of 25 is $\frac{5}{8}$ of how many fifths of 45? 40?
5. $\frac{5}{6}$ of 36 is $\frac{7}{10}$ of how many ninths of 54? 45?
6. $\frac{3}{8}$ of 40 is $\frac{5}{9}$ of how many ninths of 63? 54?
7. $\frac{11}{6}$ of 42 is $\frac{7}{8}$ of how many tenths of 80? 60?
8. $\frac{3}{10}$ of 50 is $\frac{5}{9}$ of how many thirds of 36? 33?
9. $\frac{7}{12}$ of 60 is $\frac{5}{9}$ of how many elevenths of 88? 77?
10. $\frac{9}{7}$ of 63 is $\frac{9}{11}$ of how many eighths of 96? 88?
11. $\frac{5}{8}$ of 64 is $\frac{10}{9}$ of how many twelfths of 60? 84?
12. $\frac{3}{4}$ of 48 is $\frac{6}{7}$ of how many sixths of 54? 66?
13. $\frac{7}{11}$ of 77 is $\frac{7}{9}$ of how many elevenths of 99? 121?

14. $\frac{5}{9}$ of 81 is $\frac{9}{12}$ of how many sevenths of 84 ? 49 ?
15. $\frac{11}{12}$ of 108 is $\frac{9}{7}$ of how many fifths of 60 ? 45 ?

---

16. $\frac{5}{6}$ of 18 is $\frac{10}{9}$ of how many times $\frac{3}{4}$ of 12 ? 8 ?
17. $\frac{7}{8}$ of 32 is $\frac{4}{5}$ of how many times $\frac{5}{7}$ of 14 ? 7 ?
18. $\frac{8}{5}$ of 35 is $\frac{7}{3}$ of how many times $\frac{3}{11}$ of 44 ? 33 ?
19. $\frac{3}{10}$ of 100 is $\frac{6}{12}$ of how many times $\frac{2}{12}$ of 48 ? 60 ?
20. $\frac{7}{12}$ of 144 is $\frac{11}{7}$ of how many times $\frac{3}{5}$ of $\frac{1}{3}$ of 30 ?

---

21. $\frac{1}{2}$ of 3 times $5\frac{1}{3}$ is $\frac{2}{7}$ of how many times $\frac{2}{3}$ of $\frac{1}{4}$ of 2 times 12 ?
22. $\frac{2}{3}$ of 5 times $6\frac{3}{5}$ is $\frac{11}{7}$ of how many times $\frac{3}{5}$ of $\frac{1}{2}$ of 4 times $7\frac{1}{2}$?
23. $\frac{5}{7}$ of 4 times $6\frac{3}{4}$ is $\frac{9}{11}$ of how many times $\frac{2}{7}$ of $\frac{1}{3}$ of 5 times $8\frac{2}{3}$?

CHAPTER VIII.

§ **30.**—The relative value of one part to another may, more readily, be explained by some visible representation as by means of an apple, or as represented below by lines, which may be drawn on the blackboard, and fully explained by the teacher.

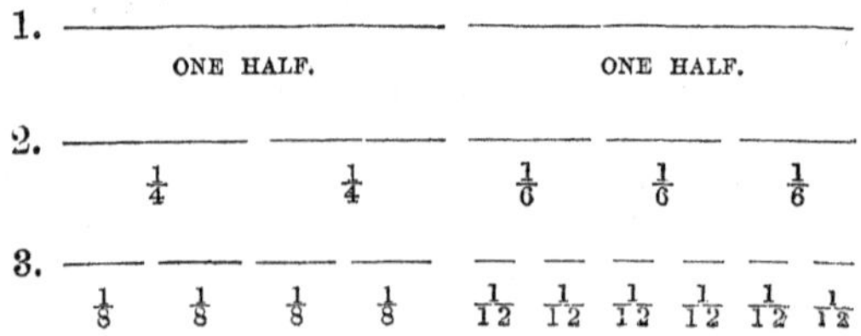

1. ———————— ————————
ONE HALF. ONE HALF.

2. ———— ———— ——— ——— ———
$\frac{1}{4}$ $\frac{1}{4}$ $\frac{1}{6}$ $\frac{1}{6}$ $\frac{1}{6}$

3. —— —— —— —— — — — — — —
$\frac{1}{8}$ $\frac{1}{8}$ $\frac{1}{8}$ $\frac{1}{8}$ $\frac{1}{12}$ $\frac{1}{12}$ $\frac{1}{12}$ $\frac{1}{12}$ $\frac{1}{12}$ $\frac{1}{12}$

Note.—A whole line may be drawn on the board; underneath draw lines one half the length, and under the lines representing halves draw lines representing eighths, and sixths, &c. It would be well also for the teacher to prepare small pieces of wood, of lengths, varying to represent the different parts of a unit.

1. If you divide an apple into 2 equal parts, and take away 1 of those parts, how much of the apple will you take away? How many halves in 1? in $1\frac{1}{2}$? in 2? $2\frac{1}{2}$? 3? $3\frac{1}{2}$? $5\frac{1}{2}$? 7? $8\frac{1}{2}$? 10? $11\frac{1}{2}$? $12\frac{1}{2}$?

2. If a yard of cloth is divided into 3 equal parts, what is one of those parts called? How many thirds in 1? $2\frac{1}{3}$? $5\frac{2}{3}$? $7\frac{1}{3}$? $12\frac{2}{3}$?

3. If you divide an apple into 4 equal parts, and take away one of those parts, how many will be left? What part of the whole apple? How many fourths in 1? in 3? 8? $9\frac{3}{4}$? $10\frac{1}{4}$? $11\frac{2}{4}$?

Note.—The number of parts should be written by the pupils on the black-board, so as to be seen by the entire class. Two parallel horizontal lines are the sign of equality; thus $=$. It shows that the numbers between which it is placed are equal to each other; as $3 = \frac{12}{4}$, and is read 3 is equal to $\frac{12}{4}$, or 3 equals $\frac{12}{4}$.

4. If a piece of cloth be divided into 5 equal parts, what will 1 of those parts be called? 2 parts? 3 parts? How many are equal to a unit?

5. If a number, or an apple, be divided into 6 equal parts, what will one of those parts be called? How many sixths $= 1$?

6. If you had $\frac{1}{2}$ of an apple, and should divide it into 2 equal parts, what would 1 of those parts be called? What is $\frac{1}{2}$ of $\frac{1}{2}$?

7. When the numerator and denominator (see § 20) are equal, what is the value of the fraction? Answer, 1.

8. When the numerator is *greater* than the denom-

inator, what is the value of the fraction? Ans. It is greater than 1.

9. When the numerator is *less* than the denominator, what is the value of the fraction?—Ans. It is less than 1.

10. How many units in $\frac{4}{2}$?
11. How many units in $\frac{6}{3}$?
12. How many units in $\frac{8}{2}$?
13. How many units in $\frac{12}{3}$?
14. How many units in $\frac{15}{5}$?
15. How many units in $\frac{30}{6}$?
16. How do you ascertain the number of units in an improper fraction?

---

17. In $2\frac{1}{2}$ how many halves?
18. In $3\frac{1}{3}$ how many thirds?
19. In $7\frac{1}{5}$ how many fifths?
20. In $5\frac{3}{4}$ how many fourths?
21. In $4\frac{2}{3}$ how many thirds?
22. In $6\frac{1}{6}$ how many sixths?
23. In $8\frac{1}{7}$ how many sevenths?
24. In $9\frac{1}{9}$ how many ninths?
25. In $7\frac{3}{8}$ how many eighths?
26. In $11\frac{11}{12}$ how many twelfths?
27. In $9\frac{5}{7}$ how many sevenths?
28. In $12\frac{7}{11}$ how many elevenths?

Note.—Numbers like $2\frac{1}{2}$, $3\frac{1}{3}$, $7\frac{1}{5}$, &c., are called MIXED numbers.

29. How do you reduce a mixed number to an improper fraction.—Ans. Multiply the whole number by the denominator of the fraction, add the numerator to the product, and read or write the result as the numerator of the new fraction.

30. How can you determine the value of an improper fraction? Why?

§ **31.**—If both terms of the fraction be multipled or divided by the same number, the value of the fraction will not be changed; because, by the operation, the *number* of the parts are increased in the same ratio that their *size* is diminished.

§ **32.**—Any whole number may be reduced to the form of an improper fraction by multiplying it by the denominator of the fraction to which form it is required to be reduced.

1. Multiply both terms of the fraction $\frac{1}{2}$ by 2. Is there any difference in the value of $\frac{1}{2}$ of a dollar, and $\frac{2}{4}$ of a dollar? Why?
2. Change 4 to an improper fraction with 5 for its denominator.
3. Change 7 to an improper fraction, having 3 for its denominator.
4. Change 9 to an improper fraction, having 7 for its denominator.
5. Change 11 to an improper fraction, having 6 for its denominator.
6. Change 6 to an improper fraction, having 7 for its denominator.
7. Change 12 to an improper fraction, having 9 for its denominator.
8. Change 8 to an improper fraction, having 11 for its denominator.
9. Change 5 to an improper fraction, having 12 for its denominator.
10. What fraction is equal to $\frac{5}{7}$, having both terms multiplied by 5?
11. Multiply both terms of the fraction $\frac{3}{5}$ by 7; by 6; by 9.
12. Multiply both terms of the fraction $\frac{7}{8}$ by 9; by 3; by 6.

13. How many eighths in $\frac{1}{2}$, $\frac{1}{4}$, and $\frac{1}{8}$?
14. How many ninths in $\frac{2}{3}$, $\frac{4}{6}$, and $\frac{2}{9}$?
15. How many tenths in $\frac{1}{2}$, $\frac{3}{5}$, and $\frac{7}{10}$?
16. How many twelfths in $\frac{1}{2}$, $\frac{2}{3}$, $\frac{5}{6}$, $\frac{3}{4}$, and $\frac{11}{12}$?

Note.—When several fractions have the same denominator in common, they are said to have a *common denominator.* The process of finding the sum of several fractions, as in the foregoing examples, is called *adding* fractions. If the denominators are alike, they are added by adding their numerators. If they have not a common denominator, they must first be reduced to a common denominator. This change is effected by multiplying all the denominators together for a new denominator, and the numerators of the respective fractions by all the denominators except its own; or if the denominators are each contained an exact number of times in the largest denominator, they may be multiplied by a number that will make them equal to the largest denominator; care being taken to multiply both terms of the respective fractions by the same number.

17. What is the sum of $\frac{2}{3}$ and $\frac{3}{4}$? $\frac{1}{2}$ and $\frac{3}{4}$? $\frac{2}{5}$ and $\frac{7}{10}$? $\frac{4}{7}$ and $\frac{1}{2}$?

18. What is the sum of $\frac{1}{2}$, $\frac{1}{3}$ and $\frac{1}{4}$? $\frac{1}{6}$, $\frac{3}{4}$ and $\frac{7}{12}$? $\frac{2}{5}$, $\frac{1}{2}$ and $\frac{4}{10}$?

19. What is the sum of $\frac{2}{9}$ and $\frac{5}{6}$? $\frac{1}{7}$ and $\frac{3}{5}$? $\frac{2}{7}$ and $\frac{3}{5}$? $\frac{7}{8}$ and $\frac{5}{9}$?

20. James gave to one of his school-fellows $\frac{1}{8}$ of an orange, to another $\frac{3}{8}$, and to a third he gave $\frac{2}{8}$; what part of the whole orange did he give away?

Solution.—He gave away a part equal to the sum of $\frac{1}{8}$, $\frac{3}{8}$, and $\frac{2}{8}$; $\frac{1}{8}$ and $\frac{3}{8}$ are $\frac{4}{8}$, and $\frac{2}{8}$ are $\frac{6}{8}$. Therefore he gave away $\frac{6}{8}$ of the orange.

21. Mary has $\frac{2}{4}$ of an apple and Delia has $\frac{4}{16}$ of an apple; what part of an apple have they both together?

Note.—Reduce the fractions to their lowest terms. This operation is performed by dividing both terms of the fraction by any number that will divide both without any remainder. Continue so to divide the fraction until both terms cannot be divided by the same number, when the fraction will be in its lowest terms.

22. A merchant has $\frac{9}{12}$ of a barrel of sugar in 1 barrel, $\frac{15}{30}$ of a barrel in another, and $\frac{6}{7}$ of a barrel in another; how much sugar is in all the barrels?

23. A grocer sold to one man $\frac{1}{3}$ of a pound of coffee, to another man $\frac{6}{12}$ of a pound, to another $\frac{12}{36}$ of a pound, and to another $\frac{3}{16}$ of a pound; how much did he sell in all, and to which one did he sell the smallest quantity?

24. If I give one boy $\frac{1}{7}$ of an apple, and another $\frac{3}{7}$, how many sevenths will I give to both?

25. A farmer plowed $1\frac{3}{4}$ acres one day, and $1\frac{3}{8}$ acres the next; how many acres did he plow in both days?

---

§ **33.**—1. A lady bought 3 yards of calico, at $\frac{1}{4}$ of a dollar a yard; how much did it cost her?

Solution.—If 1 yard cost $\frac{1}{4}$ of a dollar, 3 yards would cost 3 times $\frac{1}{4}$ of a dollar; 3 times $\frac{1}{4}$ of a dollar is $\frac{3}{4}$ of a dollar. Therefore, it cost her $\frac{3}{4}$ of a dollar.

2. In $\frac{3}{4}$ of a dollar how many cents?

Note.—Let the value of fractional parts of a dollar be given in cents.

3. A man bought 4 gallons of molasses, at $\frac{3}{5}$ of a dollar a gallon; how much did it cost him?

4. At $\frac{2}{5}$ of a dollar a bushel for corn, what would 12 bushels cost?

5. If James receive $\frac{3}{5}$ of an apple, and Laura receive 7 times as much, how much will she receive?

6. What will be the cost of 12 pounds of butter, at $\frac{3}{11}$ of a dollar a pound?

7. If a horse eat $\frac{3}{5}$ of a bushel of oats in 1 day, how many bushels will he eat in 9 days? 10 days? 12 days?

8. If a pound of coffee cost $\frac{2}{15}$ of a dollar, what will be the cost of 9 pounds, at the same rate?

9. If $\frac{3}{4}$ of a ton of hay will last a span of horses 1 week, how much will it take to last 2 span of horses 5 weeks?

10. If $\frac{1}{3}$ of a barrel of flour will last 7 persons a week, how many barrels will last 14 persons 4 weeks?

Note.—First find how much will last 14 persons one week.

11. If $\frac{7}{8}$ of a yard of satin will make 1 vest, how many yards will it take to make 4 vests? 6? 7?

12. A farmer sold 12 bushels of potatoes, at $\frac{1}{3}$ of a dollar a bushel; how much did they come to?

13. What would have been the cost of 24 bushels, at the same rate?

14. When flour is $\frac{3}{5}$ of a dollar per hundred weight, what will be the cost of 11 hundred weight? 12 hundred weight? 13 hundred weight?

---

§ **34.**—1. If 1 yard of silk cost $1\frac{2}{3}$ dollars, what will 5 yards cost at the same rate?

Solution.—If 1 yard cost $1\frac{2}{3}$ dollars, 5 yards will cost 5 times $1\frac{2}{3}$ dollars; 5 times 1 dollar = 5 dollars, and 5 times $\frac{2}{3}$ of a dollar = $\frac{10}{3}$ of a dollar = $3\frac{1}{3}$ dollars, which added to 5 dollars, = $8\frac{1}{3}$ dollars. Therefore, 5 yards will cost $8\frac{1}{3}$ dollars.

2. What will 6 weeks' board come to at $2\frac{1}{2}$ dollars a week?

3. If 1 barrel of flour cost $3\frac{2}{5}$ dollars, what will 8 barrels cost?

4. At $4\frac{3}{7}$ dollars a yard for broadcloth, what will 9 yards come to? 5 yards? 7 yards?

5. If 1 yard of satin cost $2\frac{1}{7}$ dollars, and a yard of cloth cost $3\frac{6}{7}$ dollars, what will be the cost of 5 yards of satin and 5 yards of the other cloth?

6. What is the sum of the products of 5 times $5\frac{3}{7}$, and 5 times $7\frac{4}{7}$?

7. 5 men did a certain piece of work in $9\frac{5}{7}$ days; how long would it have taken 1 man to finish the same work?

8. A grocer sold one man 7 pounds of coffee at $9\frac{1}{2}$ cents per pound, and to another 6 pounds of sugar at $7\frac{3}{4}$ cents a pound; how much did he receive from both men?

9. A lady purchased 11 yards of silk at $\frac{7}{5}$ of a dollar a yard; how much did she pay for it?

10. What will 10 firkins of butter come to at $7\frac{2}{7}$ dollars a firkin?

11. What will $9\frac{3}{5}$ tons of hay come to at 11 dollars a ton?

12. A man bought a vest for $3\frac{1}{3}$ dollars, a coat for $7\frac{1}{8}$ dollars, and a pair of pantaloons for $2\frac{1}{3}$ dollars; how much would he have had to pay for 3 of each?

13. How many is 5 times $1\frac{1}{4}$?

14. How many is 6 times $2\frac{2}{5}$?

15. How many is 7 times $3\frac{2}{3}$?

16. How many is 8 times $4\frac{5}{7}$?

17. How many is 9 times $5\frac{3}{8}$?

18. How many is 10 times $6\frac{1}{5}$? $5\frac{3}{11}$?

19. How many is 11 times $7\frac{5}{11}$? $3\frac{5}{12}$?

20. How many is 12 times $8\frac{1}{2}$? $9\frac{3}{7}$?

21. How many are 3 times $5\frac{1}{4}$ and 4 times $\frac{3}{8}$?

22. How many are 5 times $6\frac{2}{7}$ and 6 times $5\frac{5}{6}$?

23. How many are 6 times $9\frac{1}{2}$ and 7 times $8\frac{2}{3}$?

---

§ 35.—1. A man planted $\frac{1}{5}$ of his land to corn, and $\frac{3}{5}$ to tobacco, and kept the remainder for other grain and pasture; how much was left for pasture and grain?

2. A man bought $3\frac{2}{3}$ barrels of cider and sold $1\frac{1}{3}$ barrels: how much had he remaining?

3. If a farm of 36 acres be diminished by $\frac{1}{2}$ and $\frac{1}{3}$ of the whole number of acres, how much of the farm remains?

4. A man bought $17\frac{3}{7}$ bushels of peaches, and sold to one man $1\frac{2}{7}$ bushels, and to another $5\frac{1}{7}$ bushels; how many bushels remained unsold?

5. A boarder was indebted for 7 weeks board at $2\frac{2}{3}$ dollars a week, and paid at one time $2\frac{1}{2}$ dollars, at another time $11\frac{3}{4}$ dollars; how much did he then owe?

6. What is the difference between $11\frac{1}{2}$ and $3\frac{1}{4}$?

7. If a man should give $37\frac{2}{5}$ dollars for a horse, and sell him again for 12 bushels of seed wheat at $2\frac{1}{5}$ dollars a bushel, and $5\frac{1}{5}$ dollars in money, would he gain or lose by the bargain, and how much?

8. What is the difference between $20\frac{4}{5}$ and $10\frac{2}{5}$?

9. What is the difference between $7\frac{5}{7}$ and $5\frac{3}{7}$?

10. What is the difference between $8\frac{3}{4}$ and $3\frac{1}{2}$?

11. What is the difference between $5\frac{5}{9}$ and $4\frac{1}{3}$?

12. What is the difference between $11\frac{1}{11}$ and $9\frac{1}{11}$?

13. What is the difference between $13\frac{2}{3}$ and $11\frac{1}{3}$?

14. What is the difference between $18\frac{5}{6}$ and $15\frac{1}{2}$?

15. What is the difference between $27\frac{10}{11}$ and $23\frac{9}{11}$?

16. If a man give $8\frac{3}{5}$ dollars a piece for 3 cows, and $2\frac{3}{5}$ dollars a piece for 7 calves, how much less would the calves cost than the cows?

17. How much more will 5 barrels of flour cost, at $5\frac{1}{5}$ dollars a barrel, than $7\frac{1}{5}$ bushels of wheat, at 3 dollars a bushel?

18. There is a pole standing in the water, so that $\frac{1}{2}$ is in the mud, $\frac{1}{3}$ in the air, and the balance in the water; how much is in the water?

19. A shepherd has 64 sheep in four fields; the first contains $\frac{1}{8}$, the second $\frac{1}{4}$, the third $\frac{1}{2}$ of them; what part of them are in the fourth field?

20. A lady purchased $5\frac{1}{7}$ yards of calico at one store, and $3\frac{3}{7}$ at another; she used $5\frac{2}{7}$ yards; how much had she then remaining?

21. A farmer has a quantity of grain, of which $\frac{1}{3}$ is wheat, $\frac{1}{4}$ corn, $\frac{1}{8}$ oats, $\frac{1}{12}$ barley, and the remainder rye; what part is rye? If he has 108 bushels in all, how many are there of each kind?

---

§ 36.—1. At $\frac{1}{3}$ of a dollar a bushel, how much wheat can be bought for $3\frac{2}{3}$ dollars?

2. If one pound of tea cost $\frac{2}{3}$ of a dollar, how many pounds can be bought for $5\frac{1}{3}$ dollars?

3. A mother divided 10 dollars equally among her 4 children, what was each one's share?

4. At $\frac{3}{4}$ of a dollar a piece, how many chairs can be purchased for 12 dollars? 16 dollars? 20 dollars?

SOLUTION.—As many as the number of times that $\frac{3}{4}$, the price of one chair, is contained in 12 dollars. In 12 dollars are $\frac{48}{4}$ of a dollar, and $\frac{3}{4}$ of a dollar is contained in $\frac{48}{4}$ of a dollar 16 times. Therefore, at $\frac{3}{4}$ of a dollar a piece, 16 chairs can be bought for 12 dollars.

5. If one bushel of apples cost $\frac{2}{5}$ of a dollar, how many bushels may be bought for 16 dollars? 8 dollars? 6 dollars?

6. What is $\frac{1}{2}$ of 1? 2? 3? 4? 5? 6? 7? 8? 9? 10? 11? 12? 20?

7. What is $\frac{1}{3}$ of 3? 1? 5? 8? 4? 7? 8? 9? 11? 12? 13? 18? 28?

8. What is $\frac{1}{4}$ of 1? 4? 8? 9? 12? 13? 15? 8? 21? 33? 40? 10? 19?

9. What is $\frac{1}{6}$ of 1? 3? 19? 15? 20? 23? 28? 32? 40? 50? 36? 80?

10. What is $\frac{1}{5}$ of 1? 4? 8? 10? 20? 18? 28? 38? 47?

11. What is $\frac{1}{8}$ of 1? 17? 19? 29? 37? 45? 48? 58?

12. What is $\frac{1}{10}$ of 1? 20? 33? 44? 55? 66? 77? 88? 99?

13. What is $\frac{1}{7}$ of 1? 21? 15? 25? 41? 52? 63? 71? 99?

14. What is $\frac{1}{9}$ of 1? 18? 28? 40? 37? 51? 65? 84? 90?

15. What is $\frac{1}{11}$ of 1? 23? 35? 46? 58? 67? 79? 80? 99?

16. What is $\frac{1}{12}$ of 1? 36? 38? 40? 52? 74? 83? 94? 97?

17. What is $\frac{1}{11}$ of 2? 47? 87? 97? 57? 74? 78? 75? 131?

---

§ **37.**—1. What is $\frac{1}{3}$ of 5? $\frac{2}{3}$? $\frac{3}{3}$?

SOLUTION.—$\frac{1}{3}$ of 1 is $\frac{1}{3}$; $\frac{1}{3}$ of 5 is $\frac{5}{3}$ which equals $1\frac{2}{3}$. Therefore, $1\frac{2}{3}$ is $\frac{1}{3}$ of 5.

2. What is $\frac{1}{4}$ of 12? $\frac{2}{4}$ of 12? $\frac{3}{4}$ of 12?
3. What is $\frac{1}{7}$ of 13? $\frac{2}{7}$ of 13? $\frac{5}{7}$ of 13?
4. What is $\frac{1}{9}$ of 19? $\frac{2}{9}$ of 19? $\frac{7}{9}$ of 19?
5. What is $\frac{1}{11}$ of 12? $\frac{3}{11}$ of 12? $\frac{5}{11}$ of 12?
6. What is $\frac{1}{5}$ of 6? $\frac{3}{5}$ of 6? $\frac{4}{5}$ of 6?
7. What is $\frac{1}{7}$ of 8? $\frac{2}{7}$ of 8? $\frac{6}{7}$ of 8?
8. What is $\frac{1}{8}$ of 9? $\frac{3}{8}$ of 9? $\frac{7}{8}$ of 9?
9. What is $\frac{1}{10}$ of 21? $\frac{2}{10}$ of 21? $\frac{9}{10}$ of 21?
10. What is $\frac{1}{12}$ of 60? $\frac{3}{12}$ of 60? $\frac{11}{12}$ of 60?
11. What is $\frac{1}{12}$ of 49? $\frac{5}{12}$ of 49? $\frac{2}{12}$ of 49?
12. What part of 12 is 7?
13. What part of 8 is 5?
14. 11 is how many times 3? 5? 7? 9? 11? 13? 19?
15. If 4 men can do a piece of work in five months, in what time can 6 men do it?

SOLUTION.—If 4 men can do it in 5 months, it would require 4 times 5 months for 1 man to do it; 4 times 5 months are 20 months. 6 men will do it in $\frac{1}{6}$ of 20 months; $\frac{1}{6}$ of 20 months is $3\frac{2}{6}$ months. Therefore, 6 men would do it in $3\frac{2}{6}$ months.

16. If 3 barrels of cider cost 8 dollars, what will 5 barrels cost? 13 barrels? 11 barrels?

17. If a quantity of provisions will last 7 men 10 days, how long will it last 9 men? 11 men?

18. At 3 cents for 5 apples, how many could you buy for 18 cents? 20? 30? 40? 50? 60?

19. If 5 barrels of flour cost 17 dollars, what would 7 barrels cost? 9 barrels? 11 barrels?

20. If 2 pipes will empty a hogshead of wine in 3 hours, how long will it require 4 pipes to empty it?

21. If a man can earn 5 dollars in 9 days, how much can he earn in 12 days? 10 days?

22. A merchant sold 3 yards of cloth for 16 dollars; what will 7 yards cost at the same rate?

23. If 5 barrels of flour cost 21 dollars, what will 11 barrels cost at the same rate?

24. If 5 bushels of corn cost 8 dollars, what will 3 bushels cost?

25. If 9 bushels of wheat cost 12 dollars, what will 7 bushels cost?

26. If 3 barrels of flour cost 8 dollars, what will 11 barrels cost?

27. A man had 64 dollars, and spent $\frac{7}{8}$ of it; how much had he left?

28. If 10 men can do a piece of work in 7 days, in what time could 9 men do the same work?

29. A man having a flock of 108 sheep, lost $\frac{2}{9}$ of them and afterwards sold $\frac{5}{9}$ of his original flock for $2\frac{1}{3}$ dollars a head; how many did he lose, what number did he sell, and for how much, and what number had he remaining?

30. If 4 pipes will fill a cistern in 9 hours, in what time will 7 pipes fill it? 3 pipes? 9 pipes? 11 pipes?

---

§ 38.—1. If $\frac{1}{2}$ a yard of cloth cost $1\frac{1}{2}$ dollars, what would 1 yard cost?

2. $1\frac{1}{2}$ is $\frac{1}{2}$ of what number?—Ans. $1\frac{1}{2}$ is $\frac{1}{2}$ of 2 times $1\frac{1}{2}$; 2 times 1 is 2, and 2 times $\frac{1}{2}$ is $\frac{2}{2} = 1$, which added to 2 is 3.

3. If $\frac{1}{5}$ of a box of lemons cost $3\frac{2}{3}$ dollars, what will a whole box cost?

4. $3\frac{2}{3}$ is $\frac{1}{5}$ of what number? (*a*)
5. $4\frac{5}{9}$ is $\frac{1}{7}$ of what number?
6. $3\frac{1}{8}$ is $\frac{1}{5}$ of what number?
7. $5\frac{3}{7}$ is $\frac{1}{4}$ of what number?
8. $7\frac{3}{4}$ is $\frac{1}{5}$ of what number?
9. $6\frac{7}{10}$ is $\frac{1}{7}$ of what number?
10. $8\frac{3}{4}$ is $\frac{1}{8}$ of what number?
11. $7\frac{1}{11}$ is $\frac{1}{7}$ of what number?
12. $9\frac{4}{9}$ is $\frac{1}{10}$ of what number?
13. $8\frac{5}{7}$ is $\frac{1}{9}$ of what number?
14. $10\frac{5}{8}$ is $\frac{1}{10}$ of what number?
15. $8\frac{1}{2}$ is $\frac{1}{11}$ of what number?
16. $12\frac{1}{12}$ is $\frac{1}{9}$ of what number?
17. $11\frac{4}{12}$ is $\frac{1}{12}$ of what number?

(*a*) Solution.—If $3\frac{2}{3}$ is $\frac{1}{5}$, 5 times $3\frac{2}{3}$ is $\frac{5}{5}$; (or the whole number;) 5 times 3 is 15, and 5 times $\frac{2}{3}$ is $\frac{10}{3} = 3\frac{1}{3}$, which added to 15 makes $18\frac{1}{3}$. Therefore, $3\frac{2}{3}$ is $\frac{1}{5}$ of $18\frac{1}{3}$.

18. A man bought a barrel of flour, paying at the rate of $3\frac{1}{11}$ dollars for each fifth; what did it cost him?

---

§ 39.—1. If 2 thirds of a bushel of wheat cost 5 shillings, what would $\frac{1}{3}$ of a bushel cost? 5 is 2 times what number?

SOLUTION.—If $\frac{2}{3}$ of a bushel cost 5 shillings, $\frac{1}{3}$ of a bushel would cost $\frac{1}{2}$ of 5 shillings; $\frac{1}{2}$ of 5 shillings is $2\frac{1}{2}$ shillings. Therefore, $\frac{1}{3}$ of a bushel would cost $2\frac{1}{2}$ shillings. 5 is 2 times $\frac{1}{2}$ of 5; $\frac{1}{2}$ of 5 is $2\frac{1}{2}$; therefore, 5 is 2 times $2\frac{1}{2}$.

2. A man paid 14 dollars for $\frac{5}{9}$ of a cask of wine; what would $\frac{1}{9}$ have cost at the same rate?

3. 16 is $\frac{7}{5}$ of some number; what is $\frac{1}{5}$ of the same number?

4. 5 is 3 times what number?
5. 7 is 4 times what number?
6. 6 is 5 times what number?
7. 9 is 7 times what number?
8. 8 is 9 times what number?
9. 11 is 6 times what number?
10. 10 is 8 times what number?
11. 12 is 11 times what number?
12. 13 is 7 times what number?
13. 15 is 8 times what number?
14. 20 is 3 times what number?
15. 28 is 9 times what number?
16. 37 is 6 times what number?
17. 39 is 7 times what number?
18. 47 is 8 times what number?
19. 53 is 11 times what number?
20. 67 is 5 times what number?
21. 89 is 7 times what number?

22. If a man can travel $1\frac{1}{5}$ miles in $\frac{1}{12}$ of a day, how far can he travel in 1 day? in 3 days? in 7 days?

23. If I pay $3\frac{1}{7}$ dollars for $\frac{1}{5}$ of a barrel of trout, what would be the cost of a whole barrel? of 3 barrels? of 8 barrels?

24. If $3\frac{1}{7}$ cords of wood cost $\frac{1}{5}$ of a dollar, what would 5 cords cost?

25. Bought $\frac{7}{5}$ of a ton of hay for 11 dollars; what would 2 tons cost at the same rate?

26. If $\frac{2}{3}$ of a barrel of fish cost 7 dollars, what would a barrel cost?

27. 2 is $\frac{2}{4}$ of what number?

28. 3 is $\frac{5}{7}$ of what number?

29. 5 is $\frac{3}{4}$ of what number?

30. 7 is $\frac{2}{5}$ of what number?

31. 9 is $\frac{1}{2}$ of what number?

32. 13 is $\frac{2}{9}$ of what number?

33. 15 is $\frac{4}{7}$ of what number?

34. 23 is $\frac{5}{8}$ of what number?

35. 27 is $\frac{4}{9}$ of what number?

36. 37 is $\frac{10}{11}$ of what number?

37. 48 is $\frac{5}{7}$ of what number?

38. 58 is $\frac{3}{11}$ of what number?

39. 40 is $\frac{11}{12}$ of what number?

40. 63 is $\frac{7}{8}$ of what number?

41. If you pay 41 dollars for $1\frac{3}{5}$ barrels of sugar, what is that a barrel?

Solution.—$1\frac{3}{5} = \frac{8}{5}$. If $\frac{8}{5}$ of a barrel cost 41 dollars, $\frac{1}{5}$ would cost $\frac{1}{8}$ of 41 dollars. $\frac{1}{8}$ of 41 dollars is $5\frac{1}{8}$ dollars; if $5\frac{1}{8}$ dollars is $\frac{1}{5}$, 5 times $5\frac{1}{8}$ dollars are $\frac{5}{5}$. 5 times 5 dollars are 25 dollars, and 5 times $\frac{1}{8}$ of a dollar is $\frac{5}{8}$ of a dollar, which added to 25 dollars, makes $25\frac{5}{8}$ dollars. $\frac{1}{8}$ of a dollar is $12\frac{1}{2}$ cents, and 5 eighths are 5 times $12\frac{1}{2}$ cents; 5 times 12 cents are 60 cents, and 5 times $\frac{1}{2}$ of a cent is $\frac{5}{2} = 2\frac{1}{2}$ cents, which added to 60 cents, make $62\frac{1}{2}$ cents. Therefore, 1 barrel of sugar would cost 25 dollars and $62\frac{1}{2}$ cents.

42. At 45 dollars for $1\frac{7}{8}$ barrels of oil, what would one barrel cost? 3 barrels?

43. How many times is 12 contained in $\frac{2}{5}$ of 47?

44. How many times is 11 contained in $\frac{3}{7}$ of 45?

45. A boy having 50 cents, spent $\frac{5}{7}$ of it for raisins,

at 8 cents a pound; how many pounds of raisins did he buy?

SOLUTION.—First find how much money he spent. $\frac{1}{7}$ of 50 cents is $7\frac{1}{7}$ cents; 5 times $7\frac{1}{7}$ cents are $35\frac{5}{7}$ cents; he bought as many pounds of raisins as the number of times that 8 cents, the price of 1 pound, are contained in $35\frac{5}{7}$ cents. 8 cents in 35 cents $4\frac{3}{8}$ times, and 8 in $\frac{5}{7}$ is contained $\frac{5}{56}$ of a time. $\frac{3}{8} = \frac{21}{56}$; and $\frac{5}{56}$ added, make $\frac{26}{56} = \frac{13}{28}$, which added to 4, makes $4\frac{13}{28}$. Therefore, he bought $4\frac{13}{28}$ pounds of raisins.

46. If a man can buy 7 sheep for 28 dollars, how many dollars would it require to buy 11 sheep at the same rate?

47. If $\frac{3}{4}$ of a ton of hay cost 10 dollars, what should be paid for 3 tons at the same rate?

48. What would be the cost of 4 yards of cloth, if $\frac{2}{5}$ of a yard cost $2\frac{2}{3}$ dollars?

49. Allowing $\frac{3}{5}$ of an acre of ground to produce $12\frac{2}{5}$ bushels, what would 4 acres produce, at the same rate?

50. If $9\frac{1}{2}$ tons of hay be bought for 38 dollars, what would be the cost of $5\frac{1}{2}$ tons, at the same rate?

SOLUTION.—$9\frac{1}{2} = \frac{19}{2}$; if $\frac{19}{2}$ cost 38 dollars, $\frac{1}{2}$ will cost $\frac{1}{19}$ of 38 dollars, which is 2 dollars; 2 halves, or 1 ton, will cost 2 times 2 dollars; 2 times 2 dollars = 4 dollars; $5\frac{1}{2}$ tons will cost $5\frac{1}{2}$ times 4 dollars; 5 times 4 dollars are 20 dollars, and $\frac{1}{2}$ of 4 dollars is 2 dollars, which added to 20 dollars = 22 dollars. Therefore, if $9\frac{3}{4}$ tons of hay cost 38 dollars, $5\frac{1}{2}$ tons would cost 22 dollars.

51. At 24 dollars for $5\frac{1}{2}$ barrels of flour, what would be the cost of 7 barrels?

52. A farmer gave 14 dollars for $3\frac{1}{2}$ tons of plaster,

and sold $1\frac{1}{4}$ tons to his neighbor, at the same price per ton; what did the part sold come to?

53. If 20 cords of wood can be bought for 25 dollars, what number of cords could be bought for 12 dollars?

54. What is the price per yard of cloth, when $3\frac{1}{2}$ yards cost $4\frac{3}{8}$ dollars? What would 5 yards cost?

55. If $\frac{3}{4}$ of an acre of land cost $7\frac{9}{10}$ dollars, what would be the cost of 5 acres?

56. If $2\frac{1}{3}$ tons of lead cost 30 dollars, what would be the cost of $3\frac{1}{2}$ tons?

57. When flour is $3\frac{2}{5}$ dollars a barrel, how many barrels can be bought for 7 dollars?

---

§ 40.—1. A boy had $\frac{3}{4}$ of an apple, which he wished to divide equally between his 2 sisters; what part of the whole apple did each receive?

SOLUTION.—Since he gave $\frac{3}{4}$ of an apple to two persons, he gave each $\frac{1}{2}$ of $\frac{3}{4}$; $\frac{1}{2}$ of $\frac{1}{4}$ is $\frac{1}{8}$; $\frac{1}{2}$ of $\frac{3}{4}$ is $\frac{3}{8}$; (or 3 times as much as $\frac{1}{2}$ of $\frac{1}{4}$.)

2. What is $\frac{1}{2}$ of $\frac{3}{5}$? $\frac{1}{2}$ of $\frac{4}{7}$? $\frac{1}{2}$ of $\frac{5}{6}$?
3. What is $\frac{1}{3}$ of $\frac{2}{5}$? $\frac{1}{3}$ of $\frac{5}{7}$? $\frac{1}{3}$ of $\frac{4}{5}$?
4. What is $\frac{1}{4}$ of $\frac{4}{7}$? $\frac{1}{4}$ of $\frac{6}{9}$? $\frac{1}{4}$ of $\frac{7}{8}$?
5. What is $\frac{1}{5}$ of $\frac{3}{4}$? $\frac{1}{5}$ of $\frac{2}{9}$? $\frac{1}{5}$ of $\frac{5}{6}$?
6. What is $\frac{1}{6}$ of $\frac{2}{7}$? $\frac{1}{6}$ of $\frac{5}{6}$? $\frac{1}{6}$ of $\frac{2}{3}$?
7. What is $\frac{1}{7}$ of $\frac{3}{5}$? $\frac{1}{7}$ of $\frac{2}{3}$? $\frac{1}{7}$ of $\frac{2}{7}$?
8. What is $\frac{1}{8}$ of $\frac{2}{3}$? $\frac{1}{8}$ of $\frac{1}{2}$? $\frac{1}{8}$ of $\frac{3}{5}$?
9. What is $\frac{1}{9}$ of $\frac{10}{7}$? $\frac{1}{9}$ of $\frac{1}{4}$? $\frac{1}{9}$ of $1\frac{1}{2}$?
10. What is $\frac{1}{10}$ of $\frac{3}{4}$? $\frac{1}{10}$ of $\frac{3}{5}$? $\frac{1}{10}$ of $2\frac{2}{5}$?
11. What is $\frac{1}{11}$ of $\frac{8}{9}$? $\frac{1}{11}$ of $\frac{2}{5}$? $\frac{1}{11}$ of $3\frac{1}{3}$?
12. What is $\frac{1}{12}$ of $\frac{9}{11}$? $\frac{1}{12}$ of $\frac{3}{7}$? $\frac{1}{12}$ of $4\frac{1}{4}$?

---

13. What is $\frac{2}{3}$ of $\frac{1}{2}$? $\frac{3}{4}$ of $\frac{2}{3}$? $\frac{2}{5}$ of $\frac{3}{4}$?
14. What is $\frac{5}{6}$ of $\frac{3}{4}$? $\frac{3}{7}$ of $\frac{5}{6}$? $\frac{4}{7}$ of $\frac{8}{9}$?

15. What is $\frac{5}{8}$ of $\frac{3}{7}$? $\frac{2}{9}$ of $\frac{11}{3}$? $\frac{5}{9}$ of $\frac{3}{7}$?
16. What is $\frac{7}{10}$ of $\frac{5}{7}$? $\frac{9}{10}$ of $\frac{7}{8}$? $\frac{3}{11}$ of $\frac{9}{12}$?
17. What is $\frac{5}{12}$ of $\frac{4}{5}$? $\frac{7}{12}$ of $\frac{9}{11}$? $\frac{11}{12}$ of $\frac{7}{12}$?

---

18. A person owned $\frac{1}{3}$ of a ship, and sold $\frac{1}{2}$ of his share; what part of the whole ship did he sell?

SOLUTION.—$\frac{1}{2}$ of $\frac{1}{3}$ is $\frac{1}{6}$. Hence he sold $\frac{1}{6}$ of the whole ship.

19. If 3 barrels of cider cost $4\frac{1}{3}$ dollars, what will 2 barrels cost?

SOLUTION.—If 3 barrels cost $4\frac{1}{3}$ dollars, 1 barrel will cost $\frac{1}{3}$ of $4\frac{1}{3}$ dollars; $\frac{1}{3}$ of $4\frac{1}{3}$ dollars is $\frac{13}{9}$ dollars, $= 1\frac{4}{9}$ dollars; 2 barrels will cost 2 times $1\frac{4}{9}$ dollars; 2 times 1 dollar is 2 dollars, and 2 times $\frac{4}{9}$ of a dollar is $\frac{8}{9}$ of a dollar, which added to 2 dollars $= 2\frac{8}{9}$ dollars.

20. If a man can travel $4\frac{2}{7}$ miles in 2 hours, how far can he travel in 5 hours? 6 hours? 7 hours?

21. If a staff 5 feet long cast a shadow 4 feet at 9 o'clock in the morning, what is the height of a tree that casts a shadow 84 feet at the same time?

22. If a cask of wine cost 63 dollars, what would $\frac{4}{7}$ of it come to at the same rate?

23. When the days are 12 hours long a man can perform a certain journey in 7 days; in how many days can he perform the same journey when the days are only 11 hours long?

24. What number added to $\frac{5}{7}$ of 15 will make the number 12?

25. What number added to $\frac{5}{9}$ of 40 will make the number 36?

26. 7 pounds of sugar, at 6 cents a pound, are equal to how many pounds at 11 cents a pound?

27. If you buy 12 bushels of wheat at 7 shillings a bushel, how many bushels of corn, at 5 shillings a bushel, will it take to pay for it?

28. If cloth that is 3 quarters wide is worth 6 dollars a yard, what is 5 yards of the same kind of cloth worth that is 5 quarters wide?

29. A man bought a cask of vinegar for 84 dollars; $\frac{3}{7}$ of it leaked away, and he sold the remainder for as much as the whole cost him; if none had leaked away, and he had sold the whole at the same rate as what was left in the cask, how much would he have gained by the bargain?

---

30. A man sold $1\frac{3}{5}$ gallons of vinegar for 50 cents; at the same rate what would have been the cost of 7 gallons?

Note.—50 cents is equal to $\frac{1}{2}$ a dollar.

31. If 1 horse consume $5\frac{2}{7}$ bushels of oats in 2 weeks, how much would 3 horses consume in 5 weeks?

32. If it require $3\frac{1}{5}$ yards of cloth to make 1 coat, how many yards will be required for 9 coats?

33. Two travelers are 100 miles apart, and are traveling toward each other, one at the rate of $7\frac{1}{2}$ miles per hour, and the other $3\frac{1}{2}$ miles; in what time will they meet, and how far will each one have traveled?

34. If, when the days are $10\frac{1}{2}$ hours long, a man perform a journey in $11\frac{1}{3}$ days, how long would it take him to perform it when the days are $11\frac{1}{3}$ hours long?

35. If 7 men can do a certain piece of work in $5\frac{1}{9}$ days, in what time could 4 men do the same work?

36. 4 men can mow a certain field in $6\frac{3}{7}$ days; in what time could 8 men mow it?

37. If a barrel of flour will last 5 persons $7\frac{3}{5}$ days, how long would 3 barrels last 7 persons?

38. A garrison has provisions to last 7 days, but in an engagement that followed, $\frac{3}{5}$ of the entire number were slain; how long could the remaining portion subsist on the provisions?

39. What number is that which if it be increased by its fifth, and seventh, and 23 more, will be doubled?

CHAPTER IX.

§ **41.**—SIMPLE NUMBERS are those which express things of the same kind:—as 1, 5, 8; also, 1 bushel, 4 yards, &c.

COMPOUND NUMBERS are those which express more than one kind; as the denominations of *money*, *time*, *weight*, &c.—6 dollars, 5 cents and 3 mills; 5 rods, 3 yards and 2 feet; 20 pounds, 4 shillings and 6 pence, are each compound numbers.

FEDERAL MONEY.

§ **42.**—FEDERAL MONEY is the currency of the United States. It is expressed in units according to the decimal scale of numeration, or numeration by tens.

The units of Federal Money are Eagles, Dollars, Dimes, Cents, and Mills.

10	mills (*m*)	make	1	cent, *ct.*
10	cents	"	1	dime, *d.*
10	dimes	"	1	dollar, $ or *doll.*
10	dollars	"	1	eagle, *e.*

The coins of the United States are of three kinds, viz: gold, silver, and copper.

The gold coins are the *double eagle*,* *eagle*, *half-eagle*, the quarter-eagle and dollar.*

The silver coins are the *dollar*, *half dollar*, *quarter dollar*, and the *dime* and *half dime.*

The copper coins are the *cent* and *half cent ;* than which nothing smaller is coined.

Note.—The character $ is placed before dollars or dollar's place, and, strictly speaking, designates United States Money. It was formerly written U. S., and at a later period the S was written upon the U, and finally the curve of the U was omitted and the S written upon it, thus $.

Federal money was established by Congress, August 8th, 1786. Sterling money was the chief currency in the *United States* prior to that time.

1. In 10 cents how many dimes ? in 20 cents ? 55 cents ? 84 cents ?

2. In 1 dollar how many dimes ? cents ? mills ?

3. In 35 mills how many cents ? How many cents in 40 mills ? 55 mills ? 60 mills ?

4. At 55 cents a yard, what would 2 yards of cambric cost ? How many dimes ?

5. If 1 yard of calico cost 2 d. and 5 cts., how many dimes would 12 yards cost ? How many dollars ?

6. At 75 cents a bushel, what will 2 bushels of wheat cost ? How many dimes ? dollars ?

7. What is the difference between *simple* and *compound* numbers ? What is Federal money ? How many different kinds of coin are there ? What is the largest ?—the smallest ? What ones are gold ?—what silver ?—what copper ? What was the original sign of Federal money ? When was it adopted ? What coins were last added ? What is the difference in value between a gold dollar and a silver dollar ? Which is the largest ? Why ?

---

* February 20th, 1849, Congress added the *double eagle* and *gold dollar.*

STERLING MONEY.

§ 43.—Sterling Money is the national currency of the kingdom of Great Britain.

4	farthings (*qr.*) . .	make	1 penny, *d.*
12	pence . . .	"	1 shilling, *s.*
20	shillings . . .	"	1 pound, £.
5	shillings . . .	"	1 crown, *cn.*
$4\frac{1}{5}$	crowns or 21 shillings	"	1 guinea, *ga.*

The English pound sterling is represented by a gold coin, called a *sovereign*, which, in the United States, is valued at $4.84. That is, 4 dollars 84 cents.—*Act of Congress*, 1842.

It is not clearly settled, from what this currency derived its name. By some it is said to have received its name from the Easterlings, who are supposed to be the first coiners of it; while others suppose it was so called to distinguish it from *stocks*, &c., whose value is nominal.

The *pound* is so called, from the fact, that in ancient times the silver of which it was composed weighed a pound (Troy).

The *guinea* was originally composed of gold brought from Guinea, in Africa; hence its name.

1. In 1 shilling, how many farthings? in 2 shillings?
2. How many farthings in 1 pound? 1£ 2s. 8d.?
3. How many pounds in 144 farthings? 144 pence?
4. In 1 crown how many pence? farthings?
5. If I pay 4 pence a piece for oranges, how many can I buy for 1 pound? $\frac{1}{2}$ a pound? 4s.?

AVOIRDUPOIS WEIGHT.

§ 44.—Avoirdupois Weight is used for weighing all coarse articles, groceries, &c.; as coffee, butter, sugar, hay, oats, &c., and all metals except gold and silver.

16 drams (*dr.*)	. . make	1 ounce, *oz.*
16 ounces	. . . "	1 pound, *lb.*
25 pounds	. . . "	1 quarter, *qr.*
4 quarters, or 100 lbs.	"	1 hundred weight, *cwt.*
20 cwt., or 2000 lbs.	"	1 ton, *T.*

Note.—*Gross* weight is the weight of goods in connexion with the boxes, bags, &c., which contain them. *Net* weight is the weight of the goods alone.

It was formerly the practice to reckon 28 pounds for a quarter, and 112 pounds for a hundred weight; but the practice has become nearly obsolete.

1. In 50 oz. how many pounds? in 70 oz.? in 33 oz.?
2. What part of a pound is 12 oz.? 10 oz.? 5 oz.? 25 oz.?
3. In 200 oz., how many pounds? in 150 oz.?
4. In 3 T. and 2 cwt., how many pounds?
5. If 8 oz. of sugar cost 5 cents, what will 5 pounds cost? 6 pounds? 4 lbs.? 3 lbs.? 6 oz.?
6. If 4 drams cost 2 cents, what will 1 lb. cost?
7. If 1 ton of hay cost 3 dol., what will 1 qr. cost? 3 qrs.?
8. What will 2 tons of hay come to at \$$1\frac{1}{2}$ per qr.?
9. In $\frac{1}{2}$ oz. and $\frac{1}{5}$ qr., how many drams? ounces?
10. What is the difference between 32 oz. and $\frac{3}{4}$ of a qr.?

TROY WEIGHT.

§ **45.**—Troy weight is used where great accuracy is required, as in weighing gold, silver, jewels, and generally in philosophical experiments.

24 grains, (*gr.*)	make	1 pennyweight, *dwt.* or *pwt.*
20 pennyweights	"	1 ounce, *oz.*
12 ounces	"	1 pound, *lb.*

Note.—The standard of weights and measures is different in different countries, and in different States of the American Union. The United States Government adopted a uniform standard, for the custom houses, &c., in 1834.

This weight was formerly used in weighing articles of every kind. It was introduced into Europe from Cairo, in Egypt, about the time of the Crusades in the 12th century. Its name is supposed by some to have been derived from *Troyes*, a city in France, where it was first adopted; others think it was derived from *Troy-novant*, the ancient name of *London*. (*See Thomson's Higher Arith., Art.* 250, also *Hind'sArith.*, Art. 224. Also *North American Review*, Vol. 45.)

1. For what is *Troy* weight used? Repeat the table.

2. How many pennyweights in 1 ounce? in 2 oz.? 3? 5? 9?

3. How many grains in 2 pennyweights? in 1 oz.?

4. How many ounces in 1 lb.? 2 lbs.? 4 lbs.? 5? 7?

5. How many pounds in 48 ounces? in 28? 37? 56?

APOTHECARIES' WEIGHT.

§ **46.**—Apothecaries' weight is used in compounding medicines; but in buying and selling the same, *Avoirdupois* weight is used.

20 grains, (*gr.*)	make	1 scruple, *sc.* or ℈.
3 scruples	"	1 dram, *dr.* or ʒ.
8 drams	"	1 ounce, *oz.* or ℥.
12 ounces	"	1 pound, ℔.

Note.—The *pound* and *ounce* of Troy weight, are the same as in this; the other denominations differ.

1. In 10 pounds, how many ounces? in 5 pounds? in 15 pounds?

2. In 2 lbs. 2 oz. and 4 drams, how many scruples? grains?

3. Change or reduce 144 oz. to pounds. 50 oz. 75 oz.

4. Reduce 200 drams to pounds; to ounces.

5. Reduce 1 pound to grains. 2 pounds.

DRY MEASURE.

§ 47.—Dry measure is used in measuring *grain, fruit, salt, coal,* &c.

2 pints, (*pts.*)	make	1 quart, *qt.*
8 quarts	"	1 peck, *pk.*
4 pecks	"	1 bushel, *bu.*
8 bushels . . .	"	1 quarter, *qr.*
32 bushels . . .	"	1 chaldron, *ch.*

Note.—A chaldron of coal in England is 36 bushels.

1. For what is *dry* measure used? Repeat the table.
2. In 1 bushel how many pints?
3. In 1 ch. how many pecks? how many quarts?
4. In 1 ch. how many pecks, quarts, and pints?
5. If 13 pecks of wheat cost 14 shillings, how many dollars would 6 bushels cost?

Note.—When no number of shillings is spoken of a dollar in this work, 8 is understood.

CLOTH MEASURE.

§ 48.—Cloth measure is used for measuring all goods sold by the yard; as cloth, muslin, &c.

$2\frac{1}{4}$	inches (*in*)	make	1 nail, *na.*
4	nails	"	1 quarter of a yard, *qr.*
4	quarters	"	1 yard, *yd.*
3	quarters	"	1 ell Flemish, *e. Fl.*
5	quarters	"	1 ell English, *e. E.*
6	quarters	"	1 ell French, *e. F.*

Note.—Cloth measure may be considered a species of long measure (see § 48). In measuring cloths, sold by the yard, no regard is had to the width.

1. In 2 quarters how many nails? in 3? 5? 6?
2. In 3 yards how many quarters? in 4? 6? 9?
3. In 4 ells Flemish how many yards? quarters?
4. In 5 ells English how many yards? quarters?
5. How many inches are there in 1 yard? 2 yards?
6. At 5 dollars a yard, what would be the cost of 3 ells French?

LINEAR MEASURE.

§ 49.—LINEAR or Long Measure is used in measuring lines; as *length, distance, width, height*, &c.

12	inches (*in.*)	make	1 foot, *ft.*
3	feet	"	1 yard, *yd.*
$5\frac{1}{2}$	yards	"	1 rod or pole, *r.* or *p.*
40	rods	"	1 furlong, *fur.*
8	furlongs	"	1 mile, *m.*
3	miles	"	1 league, *l.*
60 $69\frac{1}{2}$	geographic, or statute miles	"	1 degree, *deg.* or (°).
360	degrees	"	1 circle.

Also, 4 inches make 1 hand; 9 inches, 1 span; 18 inches, 1 cubit; 6 feet 1 fathom.

Note.—In measuring land, roads, &c., surveyors use a chain which is 4 rods in length, and contains 100 links. This *chain* is called *Gunter's chain*, from its inventor. The standard unit of length, adopted in the United States, is the *yard* of 3 feet, or 36 inches. It is made of brass, at the temperature of 62° Fahrenheit.

1. How many inches in 2 feet? 3? 4? 5? 6? 7?
2. How many degrees in the circumference of the earth?
3. How many feet tall is a horse that measures 16 hands?
4. In $3\frac{1}{4}$ furlongs, how many rods? in 5 furlongs?
5. In $4\frac{5}{6}$ feet, how many inches? $5\frac{3}{8}$ feet?
6. In $3\frac{1}{2}$ miles, how many rods? furlongs?

SQUARE MEASURE.

§ 50.—Square Measure is used in the measurement of surfaces, or all extension where length and breadth are both considered, without regard to thickness.

144	square inches (*sq. in.*)	make	1 square foot, *sq. ft.*
9	square feet	"	1 square yard, *sq. yd.*
$30\frac{1}{4}$	square yards	"	1 *sq.* rod, or pole, *P.*
40	square *r.* or *p.*	"	1 rood, *R.*
4	roods, or 160 sq. r.	"	1 acre, *A.*
640	acres	"	1 *sq.* mile, *M.*

A section of land contains 640 acres, and is bounded by four equal sides of a mile each; 6 miles square, or 36 square miles, make a township.

Note.—A square is a figure, bounded by four equal sides, whose angles are *right angles.*

A square foot is a square, whose sides are one foot in length.

A square yard is a figure, whose sides are each one yard, or 3 feet in length.

CUBIC MEASURE.

§ 51.—Cubic Measure is used in measuring solid bodies, or bodies in which length, breadth, and thickness are considered.

1728 cubic inches (*cu. in.*)	make	1 cubic foot, *cu. ft.*
27 cubic feet	"	1 cubic yard, *cu. yd.*
128 cubic feet	"	1 cord.

A cord of wood is a pile *four* feet wide, *four* feet high, and *eight* feet long. One foot in length of such a pile is called a cord-foot; hence, 8 *cord-feet* make 1 cord.

50 cubic feet of timber are allowed to weigh a *ton*. Of round timber, such a quantity is allowed for a ton as, when hewn, will make 40 cubic feet.—*See Dodd's Arith.*, § 167.

CIRCULAR MEASURE.

§ 52.—Circular Measure is used in measuring any part of the circumference of a circle, in reckoning *latitude* and *longitude*, and in computing the motions of the heavenly bodies.

60 seconds (″)	make	1 minute, (′).
60 minutes	"	1 degree, (°).
30 degrees	"	1 sign, *s.*
12 signs	"	1 circle, *c.*

Note.—This measure is sometimes called *Angular Measure.* The *circumference* of every circle is divided or supposed to be divided into 360 equal parts, which are called degrees.

Every degree is $\frac{1}{360}$ part of the circumference; hence, its length must depend on the size of the circle.

Illustration.—In the subjoined figure, intended to illustrate the divisions of a circle and its circumference, from A to B is 45°; from A to C 90°; from A to D is 180°; from A to E by B, C, and D, is 270°; from A to A, round the entire circle, is 360°.

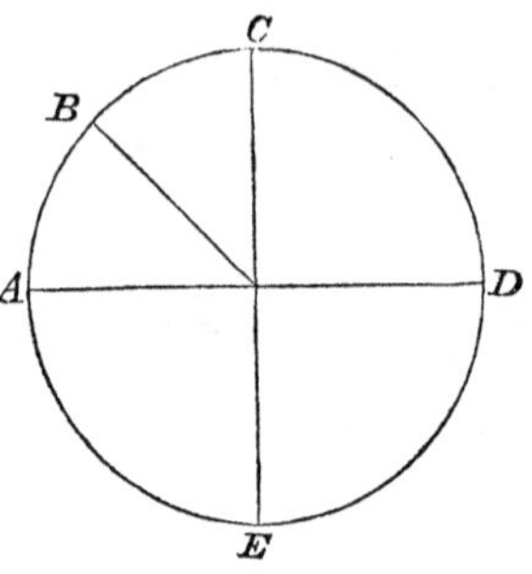

Note.—The origin of the division of the circle into 360 equal parts, was from the supposed length of the year, the number of days being reckoned in round numbers at 360, or 12 months of 30 days each. The 12 s. correspond to the 12 months.

Minutes is derived from the Latin *minutum*, which signifies a *small part.* *Seconds* signifies *second* minutes, a second order of minutes, and is an abbreviation of that phrase.

The earth turns on its axis from *west* to *east* once every 24 hours; hence it revolves 15° in *one* hour, and in 4 minutes it would revolve 1°; also in 4 seconds of time it would revolve 1′. From this we can readily tell the difference in time between any given places if their difference of longitude is known.

As the earth turns from west to east, it is clear that as we go east from any given point, we shall find their time faster, or, as it is called, *earlier time;* and slower as we go west.

1. Suppose the difference of longitude between Milwaukee and Dubuque to be 7° 30′, what is the difference in their time?

2. If it is 12 o'clock, M., at Dubuque, what is the hour at Milwaukee?

3. Is the time then at Milwaukee faster or slower than the time at Dubuque? *Why?*

4. What do you understand by faster time?

5. If it is 10 o'clock, A. M., at the Capitol in Mad-

ison, and at a place 15° from it the time is 9 o'clock, A. M., is the time at the latter place *earlier* or *later* time?

6. Is the place east or west of Madison?
7. For what is circular measure used?
8. When the difference in longitude between two places is 1′, what is the difference of time?
9. Which way does the earth revolve on its axis?
10. In what time does it perform one revolution?
11. On what does the length of a degree depend?

MEASURE OF TIME.

§ 53.—Time is naturally measured in *days* by the turning of the earth on its axis, and in *years* by its revolution around the sun.

60 seconds (*sec.*)	make	1	minute, *min.*
60 minutes	"	1	hour, *hr.*
24 hours	"	1	day, *d.*
7 days	"	1	week, *wk.*
4 weeks	"	1	lunar month, *mo.*
13 lunar months, or 52 weeks, (nearly)	"	1	year, *yr.*
12 calendar months, or 365 ds. 6 hrs., (nearly,)	"	1	civil year, *yr.*

A *solar year* is the time it requires for the earth to perform one revolution round the sun, and contains 365 ds. 5 hrs. 48 minutes and 49.6 seconds.

A *common* year contains 365 days; a *leap* year contains 366 days. 100 years make a century.

The names of the 12 calendar months, into which the common year is divided, with their number, and the number of days in each is represented below.

Name of Month.	*Abbreviation.*	*Number.*	*No. of days.*
January,	Jan.	*first,*	31
February,	Feb.	*second,*	28
March,	Mar.	*third,*	31
April,	Apr.	*fourth,*	30
May,	May,	*fifth,*	31
June,	June,	*sixth,*	30
July,	July,	*seventh,*	31
August,	Aug.	*eighth,*	31
September,	Sept.	*ninth,*	30
October,	Oct.	*tenth,*	31
November,	Nov.	*eleventh,*	30
December,	Dec.	*twelfth,*	31

The number of days in each may be more easily remembered by committing to memory the following lines :—

Thirty days has September,
April, June, and November;
All the rest have thirty-one,
Saving February alone,
Which has four and twenty-four,
And every fourth year one day more.

Or the fourth, eleventh, ninth, and sixth,
Have thirty days to each affixed;
And every other thirty-one,
Except the second month alone,
To which we twenty-eight assign,
Till Leap Year gives it twenty-nine.

Every year that can be divided by 4 without a remainder, is *leap year*, except centennial years, which must also be divisible by 400, or they are not leap years. The years 1800 and 1900 are not divisible by 400, consequently, in those years February has only 28 days, and they are not leap years. The year 2000 will be leap year.

Note.—January the first month, was so called, in honor of *Janus*, a *Roman deity*, who was supposed to preside over the commencement of the year.

The name of the second month is from the Latin *fubruo*, which signifies to purify by sacrifice, and is so called from the circumstance that this month was devoted to the purification of the people.

The *third* month, March, formerly the *first* month of the Roman year, takes its name from *Mars*, the Roman god of war.

The *fourth* month, April, is from the Latin *aperio*, which signifies *to open.* The month was so called because at this season of the year, the buds and blossoms put out.

On the first day of the fifth month, May, the ancients offered sacrifices to the goddess *Maia*, the mother of *Mercury ;* hence its name.

The sixth month, June, takes its name from the goddess *Juno.*

The seventh month, July, was so called in honor of *Julius* Cæsar, who was born in this month.

The eighth month, August, is so called in honor of Augustus Cæsar.

The ninth month, September, formerly the seventh month of the Roman year, is from the Latin *Septem*, which signifies seven.

The tenth month, October, from *Octo, eight.* It was the eighth month of the Roman year.

The eleventh month, November, is from *Novem*, nine. It was the ninth month of the Roman year.

The twelfth month, December, from *decem*, ten, was the tenth month of the Roman year.

---

§ 54.—The year is also divided further into *seasons*, as follows :—SPRING, which comprises *March, April*, and *May ;* SUMMER, which comprises *June, July*, and *August ;* AUTUMN, which comprises *September*, *October*, and *November ;* and WINTER, comprising *December*, *January*, and *February.*

---

§ 55.—The names of the days of the week, with their origin, is as follows :—

1. *Sunday*, from the *sun*. This day was dedicated to the worship of the sun, by the Ancients.

2. *Monday*, was dedicated to the worship of the moon; hence its name.

3. *Tuesday* took its name from *Tuisco*, the Saxon god of war.

4. *Wednesday* is from *Woden*, a deity.

5. *Thursday* is from *Thor*, the Danish god of storms, &c.

6. *Friday* is from *Friga*, the Saxon goddess of beauty.

7. *Saturday* is from *Saturn*, to whom it was dedicated.

WINE MEASURE.

§ **56.**—Wine measure is used in measuring all liquors except beer and ale; also for measuring molasses, oil, &c.

4	gills (*gi.*)	make	1 pint, *pt.*
2	pints	"	1 quart, *qt.*
4	quarts	"	1 gallon, *gal.*
$31\frac{1}{2}$	gallons	"	1 barrel, *bar.* or *bbl.*
2	barrels or 63 gallons	"	1 hogshead, *hhd.*
2	hogsheads	"	1 pipe or butt, *pi.*
2	pipes	"	1 ton, *tun.*

Note.—The wine gallon contains 231 cubic inches. The beer gallon 282. *Milk* is sometimes sold by one of these measures and sometimes by the other. It is proper to sell it by beer measure.

BEER MEASURE.

§ **57.**—*Beer* measure is used in measuring beer, ale, &c.

2	pints (*pts.*)	make	1 quart, *qt.*
4	quarts	"	1 gallon, *gal.*
36	gallons	"	1 barrel, *bar.* or *bbl.*
$1\frac{1}{2}$	barrels or 54 gallons	"	1 hogshead, *hhd.*

(See note, § 56.)

PAPER AND BOOKS.

§ **58.**—The number of leaves into which a sheet of paper is folded is denoted by the terms *folio*, *quarto*, &c.

24	sheets of paper	make	1 quire.
20	quires	"	1 ream.
2	reams	"	1 bundle.
5	bundles	"	1 bale.

Paper is usually sold by the quire and ream. *Printing* paper, by the ream or bundle; sometimes, however, by the pound.

A *folio* is a sheet folded in *two* leaves.
A *quarto* is a sheet folded in *four* leaves.
An *octavo* is a sheet folded in *eight* leaves.
A *duodecimo* is a sheet folded in *twelve* leaves.
A 16*mo.* is a sheet folded in sixteen leaves.
An 18*mo.* is a sheet folded in eighteen leaves.
A 32*mo.* is a sheet folded in thirty-two leaves.
A 36*mo.* is a sheet folded in thirty-six leaves.
A 48*mo.* is a sheet folded in forty-eight leaves.

Note.—The teacher should take a newspaper sheet, and fold it to the different forms before his class.

---

§ **59.**—The name given to paper indicates the size of the sheet, nearly, as will be seen by the following table.

DIMENSIONS OF ENGLISH PAPER.

Names.	*Writing.*	*Printing.*	*Drawing.*
Pott,	$15\frac{1}{2}$ by $12\frac{1}{2}$ in.		$15\frac{1}{2}$ by $12\frac{1}{2}$ in.
Small Post,	$16\frac{1}{2}$ by $13\frac{1}{2}$ in.		
Foolscap,	$16\frac{3}{4}$ by $13\frac{1}{2}$ in.		$16\frac{3}{4}$ by $13\frac{1}{2}$ in.
Crown,		20 by 15 in.	20 by 15 in.
Demy,	20 by $15\frac{1}{2}$ in.		22 by 17 in.
Medium,	$22\frac{1}{2}$ by $17\frac{1}{2}$ in.	23 by 18 in.	
Royal,	24 by $19\frac{1}{4}$ in.	26 by 20 in.	24 by $19\frac{1}{4}$ in.
Super Royal,	$27\frac{1}{2}$ by $19\frac{1}{4}$ in.		$27\frac{1}{2}$ by $19\frac{1}{4}$ in.
Imperial,	$30\frac{1}{4}$ by 22 in.		$30\frac{1}{4}$ by 22 in.

Also for printing, Double Crown, 30 by 20 in., and Double Demy, $38\frac{1}{2}$ by 26 in. There are also several varieties of Drawing paper, among which the following are most common; Atlas, 34 by $26\frac{1}{4}$ in.; Elephant, 28 by 23 in.; Double Elephant, 40 by $26\frac{3}{4}$ in.; Antiquarian, 52 by 31 in.; Double Atlas, 55 by $31\frac{1}{2}$ in.; Emperor, 68 by 48 in., &c.

Note.—Paper manufactured in the United States, is generally a trifle larger than English paper of the same name.

---

§ **60.**—In addition to the denominations included in the foregoing tables, the following are in use to a greater or less extent in different sections of the world.

12	units	make	1 dozen, *doz.*
12	dozen	"	1 gross.
12	gross	"	1 great gross.
20	units	"	1 score.
56	pounds	"	1 firkin of butter.
196	"	"	1 barrel of flour.
100	"	"	1 quintal of fish.
14	" of lead or iron	"	1 stone.
$21\frac{1}{2}$	stone	"	1 pig.
8	pigs	"	1 fother.
200	pounds	"	1 barrel of pork or beef.

§ 61.—Let the following synopsis of the tables, and aliquot or even parts be made familiar.

SYNOPSIS OF THE DENOMINATIONS IN FEDERAL MONEY.

E.		*\$.*		*d.*		*ct.*		*m.*
1	=	10	=	100	=	1,000	=	10,000
		1	=	10	=	100	=	1,000
				1	=	10	=	100
						1	=	10

ALIQUOT PARTS OF ONE DOLLAR.

50 cts.	=	$\$\frac{1}{2}$		$12\frac{1}{2}$ cts.	=	$\$\frac{1}{8}$
$33\frac{1}{3}$ cts.	=	$\$\frac{1}{3}$		10 cts.	=	$\$\frac{1}{10}$
25 cts.	=	$\$\frac{1}{4}$		$8\frac{1}{3}$ cts.	=	$\$\frac{1}{12}$
20 cts.	=	$\$\frac{1}{5}$		$6\frac{1}{4}$ cts.	=	$\$\frac{1}{16}$
$16\frac{2}{3}$ cts.	=	$\$\frac{1}{6}$		5 cts.	=	$\$\frac{1}{20}$

ALIQUOT PARTS OF \$1, NEW YORK CURRENCY.

4 shillings	=	50 cts.	=	$\$\frac{1}{2}$
2s. 8d.	=	$33\frac{1}{3}$ cts.	=	$\$\frac{1}{3}$
2 shillings	=	25 cts.	=	$\$\frac{1}{4}$
1s. 4d.	=	$16\frac{2}{3}$ cts.	=	$\$\frac{1}{6}$
1 shilling	=	$12\frac{1}{2}$ cts.	=	$\$\frac{1}{8}$
6 pence	=	$6\frac{1}{4}$ cts.	=	$\$\frac{1}{16}$

This currency is used in North Carolina, Ohio, Michigan, Wisconsin, Illinois, and in some of the other states.

ALIQUOT PARTS OF \$1 IN NEW ENGLAND CURRENCY.

3 shillings	=	50 cts.	=	$\$\frac{1}{2}$
2 shillings	=	$33\frac{1}{3}$ cts.	=	$\$\frac{1}{3}$
1s. 6d.	=	25 cts.	=	$\$\frac{1}{4}$
1 shilling	=	$16\frac{2}{3}$ cts.	=	$\$\frac{1}{6}$
9 pence	=	$12\frac{1}{2}$ cts.	=	$\$\frac{1}{8}$
6 pence	=	$8\frac{1}{3}$ cts.	=	$\$\frac{1}{12}$

Sterling Money.—Synopsis.

£		s.		d.		qr.
1	=	20	=	240	=	960
		1	=	12	=	48
				1	=	4

Aliquot Parts.

10 shillings	=	£ $\frac{1}{2}$	6 pence	=	$\frac{1}{2}$ *s.*
6s. 8d.	=	£ $\frac{1}{3}$	4 pence	=	$\frac{1}{3}$ *s.*
5 shillings	=	£ $\frac{1}{4}$	3 pence	=	$\frac{1}{4}$ *s.*
4 shillings	=	£ $\frac{1}{5}$	2 pence	=	$\frac{1}{6}$ *s.*
3s. 4d.	=	£ $\frac{1}{6}$	$1\frac{1}{2}$ pence	=	$\frac{1}{8}$ *s.*
2s. 6d.	=	£ $\frac{1}{8}$	1 penny	=	$\frac{1}{12}$ *s.*
2 shillings	=	£ $\frac{1}{10}$	1 farthing	=	$\frac{1}{4}$ *d.*
1s. 8d.	=	£ $\frac{1}{12}$	2 farthings	=	$\frac{1}{2}$ *d.*
1 shilling	=	£ $\frac{1}{20}$	3 farthings	=	$\frac{3}{4}$ *d.*

Avoirdupois Weight.—Synopsis.

T.		*cwt.*		*qr.*		*lb.*		*oz.*		*dr.*
1	=	20	=	80	=	2000	=	32000	=	512000
		1	=	4	=	100	=	1600	=	25600
				1	=	25	=	400	=	6400
						1	=	16	=	256
								1	=	16

Aliquot parts of a Pound.

8 ounces	=	$\frac{1}{2}$ *lb.*	2 ounces	=	$\frac{1}{8}$ *lb.*
4 ounces	=	$\frac{1}{4}$ *lb.*	1 ounce	=	$\frac{1}{16}$ *lb.*

Apothecaries' Weight.—Synopsis.

℔		℥		ʒ		℈		*gr.*
1	=	12	=	96	=	288	=	5760
		1	=	8	=	24	=	480
				1	=	3	=	60
						1	=	20

Dry Measure.—Synopsis.

ch.		*qr.*		*bu.*		*pk.*		*qt.*		*pt.*
1	=	4	=	32	=	128	=	1024	=	2048
		1	=	8	=	32	=	256	=	512
				1	=	4	=	32	=	64
						1	=	8	=	16
								1	=	2

Long Measure.—Synopsis.

m.		*fur.*		*rd.*		*yd.*		*ft.*		*in.*
1	=	8	=	320	=	1760	=	5280	=	63360
		1	=	40	=	220	=	660	=	7920
				1	=	$5\frac{1}{2}$	=	$16\frac{1}{2}$	=	198
						1	=	3	=	36
								1	=	12

Square Measure.—Synopsis.

A.		*R.*		*P.*		*sq. yd.*		*sq. ft.*		*sq. in.*
1	=	4	=	160	=	4840	=	43560	=	6272640
		1	=	40	=	1210	=	10890	=	1568160
				1	=	$30\frac{1}{4}$	=	$272\frac{1}{4}$	=	39204
						1	=	9	=	1296
								1	=	144

Aliquot parts of Time.

6 months	=	$\frac{1}{2}$ year.	15 days	=	$\frac{1}{2}$ month.
4 months	=	$\frac{1}{3}$ year.	10 days	=	$\frac{1}{3}$ month.
3 months	=	$\frac{1}{4}$ year.	6 days	=	$\frac{1}{5}$ month.
2 months	=	$\frac{1}{6}$ year.	5 days	=	$\frac{1}{6}$ month.
$1\frac{1}{2}$ month	=	$\frac{1}{8}$ year.	3 days	=	$\frac{1}{10}$ month.
$1\frac{1}{3}$ month	=	$\frac{1}{9}$ year.	2 days	=	$\frac{1}{15}$ month.
1 month	=	$\frac{1}{12}$ year.	1 day	=	$\frac{1}{30}$ month.

§ 62.—1. What would be the cost of 36 bushels of wheat, at 50 cents a bushel?

Note.—Had the price been \$1 a bushel, the cost would evidently have been as many dollars as there were bushels; but as the cost is only 50 cents, (\$$\frac{1}{2}$,) the cost will evidently be only one half of 36. If, therefore, we divide 36 by 2, the denominator of $\frac{1}{2}$, we shall have the true result. This is equivalent to multiplying 36 by $\frac{1}{2}$.

SOLUTION.—50 cents = \$$\frac{1}{2}$. If 1 bushel cost \$$\frac{1}{2}$, 36 bushels would cost 36 times \$$\frac{1}{2}$ = \$$\frac{36}{2}$ = \$18. Therefore, 36 bushels of wheat, at 50 cents a bushel, would cost \$18.

2. A laborer agreed to work for 8 cents an hour; how many dollars would he earn in 6 days of 12 hours each?

3. At 75 cents a yard, what would be the cost of 24 yards of linen?

4. A lady paid 33$\frac{1}{3}$ cents a yard for 9 yards of cloth; what did it come to?

5. What would be the expense of building one mile of fence at 50 cents a rod?

6. What would 48 dozen eggs cost, at 12$\frac{1}{2}$ cents a dozen?

7. What would be the difference in the cost of 28 pounds of tea, at 2 shillings a pound, if it be reckoned first in New York, and afterwards in New England currency?

8. How many shillings of New York currency are equal to 8 shillings New England currency?

9. A man bought 12 pails at 66$\frac{2}{3}$ cents each; how much did they cost?

Note.—66$\frac{2}{3}$ cents = \$$\frac{2}{3}$.

10. A man bought 144 pounds of coffee, at 12$\frac{1}{2}$ cents a pound; what did it amount to?

11. At $33\frac{1}{3}$ cents a *cwt.*, what will 1 ton of hay cost?

12. What will be the cost of 12 pounds of butter, at $16\frac{2}{3}$ cents a pound? $16\frac{2}{3}$ cents $= \$\frac{1}{6}$.

13. What will 11 bushels of wheat cost at 75 cents a bushel? at 50 cents?

14. At 20 cents each, how many back combs can be bought for 6 N. E. shillings?

15. A farmer sold 150 bushels of potatoes, at 25 cents a bushel; what did they come to?

§ 63.—1. What is a simple number? 2. A compound number? 3. What is Federal money? 4. Recite the table. 5. Of how many kinds are the coins of the United States? 6. Name the gold coins and tell their values. 7. The silver. 8. Copper. 9. When was Federal money established? 10. What is Sterling money? 11. Repeat the table. 12. By what is the Pound Sterling represented? 13. Repeat the table of Troy weight, and tell for what it is used. 14. Avoirdupois. 15. Explain the difference between gross and net weight. 16. What is the table of Linear measure? 17. How long is the school-room? 18. How wide? 19. For what is Square measure used? 20. Repeat the table. 21. Repeat the table of Cubic measure. 22. For what is it used? 23. Repeat the table of Cloth measure. 24. State the difference between Linear, Square, and Cubic measure. 25. Repeat the table of Wine measure. 26. Dry measure. 27. Time. 28. How is time divided? 29. Name the months of the year and state their origin. 30. Name the days of the week and tell their origin. 31. Name the aliquot parts of $1. 32. What are the aliquot parts of a pound Sterling?

Note.—The teacher may continue this review to any desirable extent.

CHAPTER X.

§ 65.—Per cent. and Per centage are terms used to denote a certain allowance on a hundred; that is, a part of a hundred. For example, the phrase 6 per cent. denotes $\frac{6}{100}$, 7 per cent., $\frac{7}{100}$, &c., of the sum of money, or number under consideration.

Note.—1 per cent. is written thus, .01; 2 per cent. thus, .02; 3 per cent. thus, .03; 4 per cent. thus, .04; 5 per cent. .05; 6 per cent. .06; 7 per cent. .07, &c.

Obs.—Per centage is applied to a large number of calculations in business life; as *Commission, Interest, Insurance,* &c. It is, therefore, of great importance that its principles should be fully understood by every student.

Interest.

§ 66.—Interest is the sum paid by the borrower to the lender for the use of money. It is reckoned at a certain rate *per cent., per annum;* that is, a certain number of dollars on a $100, a certain number of pounds on a £100, &c.

Note.—The terms *per cent.* and *per centage,* are from the Latin words *per* and *centum,* signifying *by the hundred.* The term *per annum,* signifies *by the year* or *for a year.*

§ 67.—The *money* on which interest is reckoned, is called the *principal.* The *per cent.* paid for the use of money is denominated the *rate.* The principal and interest added together is called the *amount.* That is, if I borrow $500 for one year, and agree to pay 10 per cent. for it, I must pay the lender, at the end of the year, $500, the sum borrowed, and $50, the interest, making the amount = $550.

Legal Interest.

§ **68.**—The rate per cent. to be paid on money is usually established by law; though it varies in different countries, and also in the different States of the American Union. In contracts, if no rate is mentioned, the lawful rate of the country where the contract is consummated, is understood.

Usurious Interest.

§ **69.**—Usury is any rate of interest *higher* than that established by law, and penalties of different kinds are attached to the laws, to which the person violating or exacting usury is liable.

---

§ **70.**—The following table exhibits the *legal rates* of interest in the different States, and the penalties attached for *usury.*

Names of States.	*Legal rates.*	*Penalty for Usury*
Alabama,	8 pr. ct.	Forfeit interest and usury.
Arkansas,	6 pr. ct.[*]	Forfeit usury.
Connecticut,	6 pr. ct.	Forfeit whole debt.
Delaware,	6 pr. ct.	Forfeit whole debt.
Florida,	8 pr. ct.	Forfeit interest and usury.
Georgia,	8 pr. ct.	Forfeit three times the usury.
Illinois,	6 pr. ct.[¶]	For't three times the usury and int.
Indiana,	6 pr. ct.	Forfeit double the usury. [due.
Iowa,	7 pr. ct.[2]	Forfeit three times the usury.
Kentucky,	6 pr. ct.	Forfeit usury and costs.
Louisiana,	5 pr. ct.[§]	Contracts exacting usury void.
Maine,	6 pr. ct.	Forfeit of the entire debt.
Maryland,	6 pr. ct.[‖]	Contracts exacting usury void.
Massachusetts,	6 pr. ct.	Forfeit three times the usury.
Michigan,	7 pr. ct.	Forfeit usury and $\frac{1}{4}$ of the debt.
Mississippi,	8 pr. ct.[†]	Forfeit usury and costs.
Missouri,	6 pr. ct.[1]	Forfeit usury and interest.
New York,	7 pr. ct.	Forfeit entire debt.
New Hampshire,	6 pr. ct.	Forfeit three times usury.
New Jersey,	6 pr. ct.	Forfeit entire debt.
North Carolina,	6 pr. ct.	Forfeit double the usury.
Ohio,	6 pr. ct.	Contracts exacting usury void.
Pennsylvania,	6 pr. ct.	Forfeit entire debt.
Rhode Island,	6 pr. ct.	Forfeit usury and interest.
South Carolina,	7 pr. ct.	Forfeit usury, interest, and costs.
Tennessee,	6 pr. ct.	Contracts exacting usury void.
Texas,	10 pr. ct.	Contracts exacting usury void.
Vermont,	6 pr. ct.	Recovery in *action* with costs.
Virginia,	6 pr. ct.	Forfeit double the usury.
Wisconsin,	7 pr. ct.[3]	
Dis't of Columbia	6 pr. ct.	Contracts exacting usury void.

* By special contract as high as 10 per cent.

¶ By contract as high as 12 per cent.

‖ 8 per cent is allowed on tobacco contracts.

† By contract as high as 10 per cent.

§ Banks are allowed 6 per cent.

1 By contract, as high as 10 per cent.

2 By agreement of parties, as high as 12 per cent.

3 Any rate is allowed that may be agreed upon. The interest paid on notes in this State ranges from 12 to 50 per cent.

Moneys due the United States draw 6 per cent. interest. In Canada, Nova Scotia, and Ireland, the legal rate of interest is 6 per cent. In France and England, it is 5 per cent.

---

§ 71.—In calculating interest, 360 days are reckoned a year; 30 days being considered a month.

1. What is the interest on $100 for one year, at 6 per cent.? at 7 per cent.? at 8 per cent?
2. A borrowed of B $200, at 7 per cent. interest, for 2 years; what was the interest due at the expiration of the time? What was the whole amount due?
3. At 6 per cent. per annum, what is the interest of $1 for 1 year? of $5? of $11? of $13? of $20? of $30?
4. If the interest of one dollar for one year is 7 cents, what will be the interest of $50 for the same time?
5. C loaned D $40, to be paid in 6 months, at 6 per cent. per annum; what did the note amount to at the end of the time?
6. At 6 per cent. per annum, what is the interest of $10 for 4 months? for 3 months? for 1 month?
7. What is the interest of $200 for 1 year and 6 months, at 6 per cent.?
8. What is the interest of $56, for 1 year and 3 months, at 7 per cent.?
9. If the interest of 2 months or 60 days is 1 per cent., what would be the per cent. for 20 days? for 30 days? for 90 days? for 15 days? for 5 days?

---

per annum. The general average on money loaned is 25 per cent., or 3 per cent. a month, where loans are made for a less time than one year.

10. What is the interest of $137, for 1 year and 3 months, at 8 per cent.?

11. A speculator borrowed $500 for 1 year at 7 per cent.; he immediately purchased a number of village lots, which he sold again at the end of 3 months for $600, on a credit of 1 year, at 10 per cent. What is the speculator's gain or loss at the end of 1 year from the time he borrowed the money?

12. What is the amount of $12, at the end of 2 years, interest being reckoned at 9 per cent.?

13. At 5 per cent., what is the amount of $56 for 1 year and 6 months?

14. At 4 per cent., what is the amount of $120 for 2 years and 8 months?

15. At 6 per cent., what is the amount of $150 for 1 year and 3 months?

16. At 30 per cent., what is the interest of $400 for 1 year?

17. What is the amount of $1, for 100 years, at 7 per cent.?

18. What is the interest of $36, for 1 yr. 3 months and 15 days, at 6 per cent.?

19. What is the interest of $45 for 1 yr. 1 mo., at 7 per cent?

20. What is the interest of $50 for 2 yrs. 2 mo., at 10 per cent.?

21. What is the interest of $200, for 4 years 6 mo., at 5 per cent.?

Note.—The teacher should continue the exercise on interest until all of the class are able to compute it readily. Let the operations be strictly mental.

Insurance.

§ **72.**—Insurance is an obligation assumed by a company or an individual, to pay for the *loss* or *dam-*

age of property by *fire*, *shipwreck*, or other casualty. *Dodd's Arith.*, § 251.

POLICY.

§ **73.**—The POLICY is the instrument of writing issued by the *insurers* to the person or persons insured.

PREMIUM.

§ **74.**—The PREMIUM is the price paid by the *insured* to the *insurer*, and is usually a *certain per cent.* on the amount of property insured, for one year, or any other specified time of risk.

1. What premium must I pay for insuring a store to the amount of $2760, at ½ per cent.?
2. A bookseller shipped a quantity of books from Buffalo to Kenosha, valued at $14,000 dollars, at 1½ per cent.; what amount of premium did he pay?
3. What is the annual insurance on a dwelling-house, valued at $2000, at 1½ per cent.?
4. What is the annual insurance on a dwelling-house, valued at $4000, at ¾ per cent.?
5. What insurance must I pay on a ship valued at $60,000, to perform an exploring expedition, if I pay 5 per cent.?
6. A man owning a schooner worth $36,000, obtained insurance upon it, at 5 per cent. for the season; what amount of premium did he pay?
7. A farmer paid $18 annually for insurance on his house, which was 2 per cent. on its value; what amount of property was covered by the policy?

Solution.—The rate of insurance being .02, it is evident that \$18 is $\frac{2}{100}$ of the amount of the property. Then if \$18 is $\frac{2}{100}$, $\frac{1}{2}$ of \$18 or \$9 is $\frac{1}{100}$; and $\frac{100}{100}$ or the whole, is $100 \times \$9 = \$900$; or $\$18 \div .02 = \$900$, the answer.

8. A merchant paid \$36 premium on goods, at 3 per cent., which were to be sent from New York to Milwaukee; what was the amount of the policy?

9. A farmer bought a cow for \$20, and offered to sell her for 5 per cent. less than he gave; how much did he ask for her?

10. A drover bought a horse for \$150, and sold him on the following day for 10 per cent. profit; how much did he sell him for?

11. A man bought a village lot for \$200, and sold it again at the end of six months, for 8 per cent. less than he gave; how much did he sell it for?

12. A man bought a farm for \$600, and sold it again for 15 per cent. advance; how much did he sell it for?

MISCELLANEOUS EXERCISES.

1. James had 9 apples, and his mother gave him 9 more; he then gave 4 to his sister and 2 to his schoolfellow; how many did he then have?

2. Sarah bought some tape for 5 cents, some thread for 11 cents, and some ribbon for 8 cents; what did the whole cost her?

3. How many are 8, 5, 6, 2, and 7?

4. How many are 5, 7, 9, 6, and 9?

5. How many are 7, 10, 8, 4, and 6?

6. How many are 10, 11, 8, 2, and 12?

7. How many are 11 and 9, less 5 and 7?

8. What is the difference in the cost of 12 yards of cloth at 12 cents a yard, and 11 pounds of coffee at 11 cents a pound?

9. What is the result of 14 increased by 11, divided by 5, multiplied by 6, diminished by 3 times 5, divided by 3, and increased by 7?

10. A young man had $20; he bought a coat for $7, a hat for $3, and a number of books for $6; how much money had he remaining?

11. A drover paid $18 for a cow, $4 for a sheep, and $8 for a hog; what did he pay for the whole, and how much more for the cow than for each of the others?

12. How many barrels of flour, at $5 a barrel, must be given in exchange for 11 yards of cloth, at $2 a yard?

13. How many barrels of salt, at $3 a barrel, can I buy for 48 bushels of wheat, at 50 cents a bushel?

Note.—The pupil will remember to consider 50 cents as $$\frac{1}{2}$.

14. 5 times 8 is how many times 1? 2? 8? 10?

15. 6 times 9 is how many times 5? 11? 7? 12?

16. 7 times 10 is how many times 3 times 4? 2 times 5?

17. 8 times $11\frac{1}{2}$ is how many thirds of 3 times 8?

18. A farmer had 5 sheep, which he sold as follows: for the first he received as many dollars, less one, as he had sheep; for the second, $\frac{1}{2}$ as much as for the first; for the third, $6, and for the fourth and fifth he received the same, which together amounted to as many dollars as the other 3; what did they all come to, and what did he receive a piece?

19. How many ounces are there in 3 pounds?

20. How many quarters are there in 9 yards?

21. In 5 quarters of a yard how many nails?

22. In $\frac{4}{5}$ of a penny, how many farthings?

23. In $\frac{3}{5}$ of a yard, how many quarters and nails?

24. In one chaldron, how many bushels? pecks?

25. In one gallon, how many pints and gills?

26. If 1 pint of oil cost 9 cents, what would 2 gallons cost at the same rate?

27. Five men bought a horse for $64; the first gave $12, the second $14, the third $20, the fourth $5 more than $\frac{1}{8}$ of the whole cost of the horse, and the fifth paid the remainder; how much did he pay?

28. A man owing his neighbor $37, paid him at one time $1 more than $\frac{1}{6}$ of the whole debt; at another time he paid $\frac{1}{2}$ of the debt then remaining; how much is he still indebted, on the supposition that what he now owes has been on interest 1 year, at 10 per cent.?

29. How much would $\frac{1}{8}$ of a hogshead of molasses cost, at $\$\frac{1}{2}$ a gallon?

30. If a man spend $28 a week, how much is that a day?

31. If your income is $50 a month, what is that a day?

32. If wine is worth 25 cents a pint, what is that a gallon?

33. If a man can earn $\$2\frac{3}{5}$ a day, what will he earn in a week? What in a month?

34. What is $\frac{5}{7}$ of a hogshead of wine worth at $64 a hogshead?

35. What part of a penny is 3 farthings?

36. What part of a shilling is 1 penny?

37. What part of a pound is 3 shillings?

38. What part of a pound is 1 shilling and 2 pence?

39. 4 shillings is what part of a pound?

40. 3 shillings is what part of a pound?

41. A farmer having 120 sheep, lost $\frac{1}{4}$ of them and sold $\frac{1}{2}$ of the remainder; he then bought 10 more; how many sheep did he then have?

42. A man bought 12 cows at \$9 a head, and sold 5 of them for \$60, and the remainder for \$11 a head; did he gain or lose by the bargain, and how much?

43. What is the difference between 8 times 8, and $\frac{1}{2}$ of 12 times 12?

44. A man bought a cow for \$$13\frac{1}{2}$ and sold her for \$$15\frac{3}{4}$; he laid out what he gained in oats at 20 cents a bushel; how many bushels did he buy?

45. If 4 acres of land produce 56 bushels of wheat, what number of bushels will 11 acres produce?

46. How many pounds of tea, at $33\frac{1}{3}$ cents a pound, can be bought for \$5?

47. If a laborer can earn \$$1\frac{1}{4}$ a day, how much can he earn in $5\frac{1}{2}$ days?

48. If 9 yards of cloth cost \$28, what will 11 yards cost at the same rate?

49. If 5 yards of calico cost \$14, what will 7 yards cost at the same rate?

50. $\frac{1}{2}$ of 48 is 3 times what number?

51. 4 times 9 is how many times 7? 5? 11?

52. A fisherman received 288 cents for fish, at 6 cents a pound; how many pounds did he sell?

53. Five men built a house for \$450; how many dollars did each man receive, the money being divided equally?

54. In a certain barrel are $17\frac{7}{10}$ gallons of wine; how many dollars will it come to at the rate of 10 cents for every tenth?

55. How can you reduce $\frac{2}{9}$ to eighteenths? Why? How $\frac{5}{12}$ to eighty-fourths? How $\frac{17}{20}$ to fortieths? How $\frac{7}{25}$ to hundredths?

56. A grocer sold to one person $\frac{1}{2}$ a pound of coffee, to another person $\frac{6}{12}$ of a pound, to another $\frac{3}{6}$ of a pound, to another $\frac{5}{8}$ of a pound, and to another $\frac{3}{18}$ of a pound; how many pounds of coffee did he sell the whole?

57. A man sold a load of wood for $\$3\frac{1}{3}$, a load of hay for $\$7\frac{2}{3}$, a firkin of butter for $\$5\frac{2}{4}$, and a cheese for $\$3\frac{3}{4}$; how much did he receive for the whole?

58. A dairyman carried 96 pounds of butter to market; he sold $10\frac{1}{2}$ pounds to one person, to another $25\frac{1}{2}$ pounds, to another $8\frac{1}{3}$ pounds, and to a fourth the rest; how many pounds did the fourth person buy?

59. A man has his sheep in 4 pastures; in the first pasture are $\frac{1}{5}$ of his whole flock, in the second $\frac{1}{2}$ of them, in the third $\frac{3}{10}$ of them, and the rest are in the fourth pasture; what part of his sheep are in the fourth pasture?

60. If a dinner for 1 man cost $\$\frac{1}{8}$, what will a dinner for 12 men cost? How many cents? How many New York shillings? How many N. E. shillings?

61. A farmer sold 8 bushels of corn for $\$3\frac{3}{10}$, 4 hogs at $\$5\frac{1}{10}$ a head, and 2 calves at $\$2\frac{1}{2}$ each; how much did he receive for the whole?

62. A man bought a horse for \$48, and sold him for $\frac{5}{4}$ of what it cost him; did he gain or lose by the bargain, and how much?

63. If 5 yards of broadcloth cost \$31, what would be the cost of 11 yards, at the same rate? 9 yards? 8 yards?

64. A drover sold a horse for \$72, and took his pay in oil at \$3 a gallon; how many gallons did he receive?

65. $\frac{4}{5}$ of 20 is how many times 6? 5? 8? 9?

66. $\frac{7}{11}$ of 77 is how many times 4? 7? 10? 12?

67. A man sold 20 cords of wood for $\$2\frac{1}{5}$ a cord, and took his pay in salt at \$4 a barrel; how many barrels of salt did he buy?

68. 8 times 9 and $\frac{5}{9}$ of 9 are how many times 5? 6? 7?

69. 4 times 15 and $\frac{1}{3}$ of 15 are how many times 4? 9? 12?

70. A man having \$$\frac{3}{5}$, gave $\frac{2}{3}$ of his money for a bushel of beans; what fraction of a dollar did the beans cost him?

71. A young lady having $\frac{4}{6}$ of a dollar, spent $\frac{3}{4}$ of it for lace; how much did the lace cost her?

72. A young man bought $3\frac{1}{2}$ yards of broadcloth, at \$$4\frac{2}{3}$ a yard; how much did it cost him?

73. If 3 bushels of corn are worth \$$4\frac{4}{5}$, how much are 5 bushels worth? 2 bushels? 7 bushels?

74. If it take $2\frac{2}{9}$ barrels of flour to last 12 persons 3 months, how much will it take to last 5 persons 4 months?

75. In how many days can 5 men do a piece of work that one man requires $17\frac{3}{4}$ days to do?

76. If $\frac{3}{5}$ of an acre of land is worth \$$3\frac{1}{2}$, how much is $\frac{1}{7}$ of an acre worth?

77. A farmer paid \$64 for a yoke of oxen, and paid $\frac{3}{4}$ of all the money he had; how much had he at first? How much remaining?

78. If I pay $37\frac{1}{5}$ cents for 4 pounds of sugar, how much must I pay for 8 pounds at the same rate?

79. If 5 sheep cost \$$15\frac{5}{6}$, how much will 7 sheep cost?

80. If 12 pounds of coffee cost \$$2\frac{1}{5}$, how much will 5 pounds cost? 3 pounds? 11 pounds?

81. A young lady earned a certain sum of money; $\frac{4}{7}$ of which she received for 11 weeks work, at \$$2\frac{1}{5}$ a week; how much money had she earned in all?

82. At \$$\frac{5}{7}$ a yard, how much cloth can be bought for \$8?

83. How many times can a cup, holding $4\frac{1}{2}$ pints, be filled from a jug holding $4\frac{1}{2}$ gallons?

84. If a class go $\frac{1}{3}$ of the way through the Arithmetic in $\frac{2}{3}$ of a month, how many weeks will it require for them to finish it?

85. How many times can a jug, holding a gallon and a half, be filled from a cask holding a barrel and a half?

86. How many yards of satin can I buy, at 5 shillings a yard, with £5 and 6 shillings?

87. A confectioner sold 14 pounds of candy for $\$3\frac{3}{5}$; how much was it a pound?

88. If 4 pounds of tobacco cost $\$\frac{3}{8}$, what is that a pound?

89. How many sheets of paper in a book of 148 pages, duodecimo size?

90. In $2\frac{1}{4}$ barrels of pork, how many pounds?

91. A woman having 10 pounds of butter, sold 5 pounds and 8 ounces; how much had she remaining?

92. In 320 inches how many yards?

93. A tree is standing by a church which is just $\frac{5}{8}$ of the height of the spire; it has been found that the ridge of the roof is 40 feet above the ground; from the ridge to the belfrey is $\frac{1}{2}$ as far as from the ground to the ridge, and the top of the spire is just as far above the belfrey, as the belfrey is above the ground; how high is the tree?

94. If a man consume 3 lbs. and 2 oz. of meat in 1 day, how much will he consume in a week? in 4 weeks?

95. If 2 men start from the same place, and travel the same way, one at the rate of $3\frac{6}{7}$ miles in an hour, and the other at the rate of $4\frac{5}{7}$ miles in an hour, how far will they be apart at the end of 5 hours? How far in 10 hours? How far in 6 days if they travel 10 hours each day?

96. Two men leave the same point and travel in opposite directions; one at the rate of 5 miles in 3 hours, and the other at the rate of 7 miles in 5 hours; how much faster did one travel than the other, and how far apart would they be after having traveled at that rate for the fourth part of a day?

97. What number is that, to which if its fifth be added, the sum will be 54 ?

98. If 30 men can perform a piece of work in 20 days, how many men will it require to finish the same work in 5 days ?

99. After a ship had been under sail 6 hours, it was found she had sailed 36 miles; how long would it take her to sail 180 miles at the same rate ?

100. A drover purchased of one man a horse, a cow, and a sheep, and being asked what he paid for each, answered, that he gave 2 times as much for the cow as for the sheep, and $\frac{1}{8}$ as much for the sheep as for the horse, and that for the whole he paid $60. What was the price of each ?

101. If a staff 2 feet long cast a shadow 5 feet long, what is the length of a pole that casts a shadow 25 feet long, at the same hour of the day ?

102. A gentleman standing by a church, observed the length of his own shadow to be 4 feet; the shadow of the church and spire at the same time measured 96 feet; knowing his own height, he readily calculated the height of the spire; his height being 6 feet, what was the height of the spire ?

103. If $80 worth of bread will furnish a garrison of 20 men 24 days, how long will $160 worth of bread serve a garrison of 30 men ?

104. A thief having 32 miles the start of an officer, makes off at the rate of 8 miles an hour; the officer pursues at the rate of 10 miles an hour; how long will it take him to overtake the thief ?

105. The number of scholars in a certain school is as follows: $\frac{1}{3}$ of all the pupils study Arithmetic, $\frac{1}{8}$ grammar, $\frac{1}{2}$ geography, and 10 learn to read; what number in each branch, and how many in the whole school ?

106. A man driving his hogs to market, was met

by another, who inquired of him where he was going with his hundred hogs; he replied he had not a hundred; but if he had, in addition to his present number, $\frac{1}{6}$ and $\frac{1}{3}$ and 2 hogs and a half, he should have a hundred; how many had he?

107. What number is that to which, if its $\frac{1}{2}$ and $\frac{1}{3}$ be added, the sum will lack 6 of being double the number?

108. A maiden being asked her age, not caring to give a direct answer, said it was 2 years more than $\frac{1}{2}$ of her mother's age; and that 12 years before that time her mother was 2 years more than half as old as her father, who was then 60 years of age. What was the maiden's age?

109. There is a tree standing in the water so that $\frac{3}{7}$ of its entire length is under the surface, and $\frac{2}{3}$ of this distance is the length from the surface of the water to the first limb, which is $12\frac{1}{2}$ feet below the top of the tree; how long is the tree?

110. If $9\frac{3}{4}$ days are required for 4 men to perform a certain piece of work, how long would it take to do it if 7 men were employed?

111. If 5 pipes will discharge the contents of a cistern in $8\frac{5}{7}$ hours, how long will it require for 7 pipes of the same size to empty it?

112. 30 men can do a piece of work in 20 days; how many men will it take to perform the same work in 8 days?

113. A farmer bought a wagon and a plow, and paid $48 cash down, which was $\frac{5}{7}$ of the price of them; what did they cost?

114. An exploring vessel having been out 18 months, the captain found that his crew had consumed $\frac{5}{7}$ of his provisions; how much longer will his provisions last, and how many months provision had he when he embarked?

115. How many barrels of flour, at \$4 a barrel, can be bought for $\frac{2}{3}$ of \$80?

116. A man bought a yoke of oxen and paid \$60, which was $\frac{2}{3}$ of the price; how many barrels of salt, at \$2$\frac{1}{2}$ a barrel, will it take to pay for them?

117. $\frac{3}{7}$ of 28 is $\frac{5}{8}$ of what number?

118. $\frac{4}{7}$ of 35 is $\frac{1}{9}$ of what number?

119. $\frac{7}{9}$ of 63 is $\frac{6}{15}$ of what number?

120. If 40 barrels of flour cost \$360, how much will 9 barrels cost?

121. If a car run 180 miles in 6 hours, how far will it run in 9$\frac{1}{4}$ hours?

122. If 4 stacks of hay will last 100 sheep 7 weeks, how long would 7 stacks last 200 sheep?

123. The yearly interest of Harriet's money, at 6 per cent., exceeds $\frac{1}{20}$ of the principal by \$100, and she does not intend to marry any man who is not scholar enough to tell her, mentally, her fortune; were you disposed to be her suitor, could you tell?

124. What is the interest of \$600 for 2yrs. 3mo. and 10ds., at 20 per cent.?

125. A said to B, give me one of your horses, and I shall have twice as many as you; and if I had one of yours, said B, we should have an equal number; how many horses had each?

CHAPTER XI.

§ 75.—In the operations thus far, we have been confined to small numbers, or to questions which have been so easily solved as to render the task comparatively easy.

In this chapter the questions involve larger and more complicated numbers. The pupil should not be content until he can solve, without hesitation, all questions mentally that involve no larger numbers than the following:

1. A man having $30, received $45 more; how many had he then?

2. A man had 61 sheep in one field, and 48 in another; how many sheep had he in both fields?

Note.—The numbers involved in this paragraph, may at first be analyzed into *tens* and *units*. In this way almost any person can readily learn to operate on large numbers, mentally; as in the following solution of the last question.

Solution.—61 is 60 and 1; 48 is 40 and 8; 60 and 40 are 100, and 1 and 8 are 9; 100 and 9 are 109. Therefore the sum of 61 and 48 is 109.

3. A man bought a wagon for 75 dollars, and a horse and harness for 84 dollars; how many dollars did they all cost?

4. If I pay 49 dollars for a wagon and 17 dollars for a harness, what will both cost?

5. What is the sum of 18 and 24?
6. What is the sum of 19 and 35?
7. What is the sum of 21 and 17?
8. What is the sum of 27 and 51?
9. What is the sum of 31 and 57?
10. What is the sum of 36 and 58?
11. What is the sum of 42 and 61?
12. What is the sum of 56 and 84?
13. What is the sum of 65 and 95?
14. What is the sum of 69 and 96?

Note.—The teacher may ask the class in turn or otherwise the following questions, the class having their books closed. The pupils should not be permitted to answer in concert until all are quick in giving the sum of any two numbers less than one hundred.

15. What is the sum of 13 and 19? of 15 and 17? of 17 and 18? of 22 and 28? of 25 and 27? of 29 and 11? of 17 and 31? of 24 and 36? of 33 and 44? of 35 and 55? of 18 and 69? of 28 and 19? of 41 and 51? of 43 and 63? of 39 and 69? of 48 and 68? of 96 and 85?

16. What is the sum of 56 and 65? of 59 and 95? of 57 and 68? of 86 and 54? of 37 and 73? of 73 and 67? of 62 and 83? of 44 and 66? of 82 and 47? of 20 and 95? of 13 and 37? of 99 and 89? of 47 and 89? of 98 and 57?

Note.—In the same manner the teacher may continue these exercises for a long time. The scholars soon become deeply interested and frequently resort to it as a pastime. (See suggestions on *different modes of teaching*, in the latter part of this work, also *Free School Journal*, of the State of Wisconsin, Vol. 1, for 1850.)

---

§ 76.—1. What will 2 pounds of coffee cost at 13 cents a pound? 3 pounds? 4 pounds? 5 pounds?

2. At 6 cents a yard, what would 13 yards of calico cost?

3. At 7 cents a yard, what would 13 yards of twist cost?

4. If sugar is 8 cents a pound, how many pounds of coffee, at 11 cents a pound, must be given in exchange for 13 pounds of sugar?

5. If tea is 9 shillings a pound, what will it cost to supply a family of 13 persons one year, on the supposition that each person makes use of one pound?

6. What must I pay for 10 cows, at $13 a head?

7. What cost 11 hogs, at $13 a head?

8. If 13 oxen are bought for $12 each, what amount of money will be required to pay for them?

9. What is the sum of 8 times 12, 3 times 8, and 2 times 13?

10. Thirteen times 13 is how many? 7 times 13? 9 times 13?

11. 2 times 14 is how many?

12. 3 times 14 is how many?

13. 4 times 14 is how many?

14. 5 times 14 is how many?

15. 6 times 14 is how many?

16. 8 times 14 is how many?

17. 9 times 14 is how many?

18. 10 times 14 is how many?

19. 11 times 14 is how many?

20. 12 times 14 is how many?

21. At 14 cents a yard, what will 13 yards of cloth cost?

22. If a ton of hay cost $14, what will 14 tons cost?

23. A farmer sold 15 tons of hay at $3 a ton; what did it amount to?

24. At 15 shillings a cord, what will be the cost of 4 cords of wood?

25. A lady purchased 15 skeins of silk, at 5 cents a skein; what did she pay for the whole?

26. 7 times 15 is how many?

27. 6 times 15 is how many?

28. 9 times 15 is how many?

29. 10 times 15 is how many?

30. 8 times 15 is how many?

31. 11 times 15 is how many?

32. 12 times 15 is how many?

33. 14 times 15 is how many?

34. 13 times 15 is how many?

35. 15 times 15 is how many?

36. At $2 a yard, what will 16 yards of broadcloth cost?

37. If 1 yard of cambric cost 3 shillings, what will 16 yards cost?

38. If you have to pay $4 for one barrel of flour,

what will you be required to pay for 16 barrels at the same rate?

39. 5 times 16 is how many?
40. 6 times 16 is how many?
41. 7 times 16 is how many?
42. 8 times 16 is how many?
43. 9 times 16 is how many?
44. 10 times 16 is how many?
45. 11 times 16 is how many?
46. 12 times 16 is how many?
47. 13 times 16 is how many?
48. 14 times 16 is how many?
49. 15 times 16 is how many?
50. 16 times 16 is how many?

---

§ 77.—1. How many dollars must I pay for 17 pounds of tea, at $2 a pound?

2. 3 times 17 are how many times 8? 3? 2?
3. 4 times 17 are how many times 5? 3? 6?
4. 5 times 17 are how many times 4? 6? 7?
5. 6 times 17 are how many times 5? 7? 4?
6. 7 times 17 are how many times 6? 8? 5?
7. 8 times 17 are how many times 7? 9? 6?
8. 9 times 17 are how many times 8? 10? 7?
9. 10 times 17 are how many times 9? 11? 8?
10. 11 times 17 are how many times 10? 12? 9?
11. 12 times 17 are how many times 8? 13? 10?
12. 13 times 17 are how many times 12? 11? 14?
13. 14 times 17 are how many times 13? 15? 8?
14. 15 times 17 are how many times 16? 11? 12?
15. 16 times 17 are how many times 17? 13? 9?
16. 17 times 17 are how many times 7? 11? 16?

§ **78.**—1. A farmer having 80 sheep, sold 50 of them; how many had he left?

2. What is the difference between 35 and 40? 36 and 42?

3. What is the difference between 40 and 85? 54 and 72?

4. What is the difference between 76 and 95? 48 and 93?

5. What is the difference between 49 and 91? 52 and 75?

6. A man having 57 dollars, paid a note of 12 dollars, and gave 9 dollars more for a cow; how many dollars had he left?

7. A farmer sold 18 bushels of wheat, at $2 a bushel, and took in part payment a calf at $5, and the balance in flour at $4 a barrel; how many barrels of flour did he receive?

8. What will 18 hogs come to at $3 a piece?

9. If flour cost $5 a barrel, how many barrels can be bought with 6 barrels of pork, at $18 a barrel?

10. At $7 an acre, what will 18 acres of land cost?

11. What will 18 horses cost at $50 a head?

12. 8 times 18 is how many times 20 less 11?

13. 9 times 18 is how many times 56 less 45?

14. 10 times 18 is how many times 84 less 75?

15. 11 times 18 is how many times 96 less 88?

16. 12 times 18 is how many times $\frac{1}{2}$ of 16?

17. 13 times 18 is how many times $\frac{1}{3}$ of 24?

18. 14 times 18 is how many times $\frac{1}{4}$ of 48?

19. 15 times 18 is how many times $\frac{1}{5}$ of 55?

20. 16 times 18 is how many times $\frac{2}{6}$ of 36?

21. 17 times 18 is how many times $\frac{2}{7}$ of 28?

22. 18 times 18 is how many times $\frac{3}{5}$ of 15?

§ 79.—1. What will 2 yards of calico come to at 19 cents a yard? What will 3 yards cost at the same rate?

2. When flour is $7 a barrel, how many barrels can be bought with 4 barrels of fish, at $19 a barrel?

3. What will 19 sheep come to, at $5 a head?

4. 6 times 19 is how many times $\frac{1}{5}$ of $\frac{1}{2}$ of 50?

5. 7 times 19 is how many times $\frac{2}{5}$ of $\frac{5}{7}$ of 35?

6. 8 times 19 is how many times $\frac{2}{9}$ of $\frac{1}{3}$ of 81?

7. 9 times 19 is how many times $\frac{1}{2}$ of $\frac{2}{3}$ of 33?

8. 10 times 19 is how many times 49 divided by 7?

9. 11 times 19 is how many times 3 times 3?

10. 12 times 19 is how many times $\frac{1}{3}$ of 2 times 9?

11. 13 times 19 is how many times $\frac{1}{8}$ of 6 times 12?

12. 14 times 19 is how many times $\frac{1}{5}$ of 4 times 10?

13. 15 times 19 is how many times $\frac{1}{7}$ of 7 times 9?

14. 16 times 19 is how many times $\frac{1}{9}$ of 92 less 11?

15. 17 times 19 is how many times $\frac{1}{10}$ of 11 times 11 less 3 times 7?

16. 18 times 19 is how many times $\frac{1}{11}$ of 12 times 12 less 23?

17. 19 times 19 is how many times $\frac{1}{12}$ of 15 times 16 less 8 times 12?

---

§ 80.—1. What is the difference between 5 times 8, and 2 times 20?

2. What is the difference between 3 times 20, and $\frac{1}{2}$ of 100?

3. What is $\frac{1}{5}$ of $\frac{1}{3}$ of 6 times 20?

4. If I exchange wheat at $3 a bushel, for 20 barrels of cider at $5 a barrel, how many bushels will I dispose of by the exchange?

5. If a man pay $20 a hundred for freight brought from New York city to Janesville, what will his bill amount to if he have 7 hundred weight brought?

6. The salary of the President of the United States is 25 thousand dollars a year; how much will he receive for his services for one term, which is 4 years?

---

§ 81.—Exercises in addition are of more common occurrence in the business transactions of life than most other rules of arithmetic, and the power of adding two columns of figures as *cents* is, if not indispensable, a very desirable qualification for the accomplished accountant.

Exercises of this character are not only highly useful to qualify one for the duties of business life, but serve also as one of the best tests of mental discipline yet discovered.

1. What is the sum of 15, 5, 3, 2, 8, 3, 7, 2, 9, 6, 7, 8, 5, 6, 11, 3, 5, 14, 1, 9, 4, 8, 3, 5, 7, 9, 8, 6, 4, 2, 10, 12, 4, 9, 7, 11, 6, 10, 8, 9, and 16?

2. What is the sum of 16, 4, 10, 6, 3, 7, 2, 5, 9, 11, 20, 4, 6, 8, 3, 7, 4, 3 times 5, 3, 8 less 5, 7, 4, 9, 12, 7, 3, 5, 4, 7, 8, 5, 4, 14, 8, 3, 7, 11, 5, 25, 8, 2, 4, 8, 6, 7, 8, 10, 12, 13, and 4 times 5?

3. What is the sum of 11, 5, 8, 3, 12, 4, 5, 20 less 7, 5, 8, 9, 3, 5, 13 less 5, 7, 16 less $8\frac{1}{2}$, $4\frac{1}{4}$, 7, $9\frac{1}{4}$, and 11 times 4?

4. What is the sum of 114, 14, 7, 4, $3\frac{1}{3}$, 3 times 7, 5, 8, 4, 3 times 5 less 10, 4, 8, 15 less 5, 18 divided by 3, 12, 5, 7, 2, 4, 8, 16, 5, 4, 7, 9, 11, 14, 3, 7, 9, and 11 times 2?

5. What is the difference between 114 and 112? 111 and 201? 18 and 45? 91 and 84? 47 and 56? 95 and 74? 115 and 48? 130 and 58? 250 and 145? 452 and 231?

§ 82.—1. The battle of Marathon was fought 490 years before the Christian era; how many years have elapsed since that event?

2. George Washington, of Va., was President of the United States for the period of 8 years; John Adams, of Mass., 4 years; Thomas Jefferson, of Va., 8 years; James Madison, of Va., 8 years; James Monroe, of Va., 8 years; John Q. Adams, of Mass., 4 years; Andrew Jackson, of Tenn., 8 years; Martin Van Buren, of N. Y., 4 years; W. H. Harrison, of Ohio, and John Tyler, of Va., together, 4 years; James Knox Polk, of Tenn., 4 years; how many years had the United States existed at the close of Polk's administration? How many years had Va. furnished the chief magistrate?

3. Washington was born in 1732, and died in 1799; how old was he when he died? How many years since his death?

4. The independence of the United States was declared in 1776; how many years since?

5. Christopher Columbus discovered the Continent in 1492; how many years is it since?

6. Benjamin Franklin died in 1790, and was 84 years of age at his death; in what year was he born?

7. How many years since the first English settlement was made at Plymouth, that event having transpired in 1620?

§ 83.—*Note.*—The following is designed simply as an illustration of a mode of conducting an exercise for the purpose of securing the attention of a class in such a manner as to prevent any common occurrence from removing it. Let us suppose there are two classes to recite in the same room, placed on opposite sides of the room and facing each other. Let two of the pupils act as monitors; the one acting as monitor of the right hand class standing with his back towards the left hand class; and the monitor of the left standing in the same manner before the class on the right. The books of the classes are closed and the monitors commence, one with the column marked A, and the other with the one marked B. The classes answering in concert. In large schools three divisions may be made. The monitor always standing on the opposite side of the room from his class.

A	B	C
How many are	How many are	How many are
1. 45 and 18?	16 and 32?	5 and 11?
2. 11 and 29?	14 and 29?	17 and 15?
3. 28 and 41?	33 and 55?	28 and 14?
4. 55 and 33?	41 and 82?	41 and 16?
5. 15 and 27?	18 and 17?	20 and 56?
6. 17 and 31?	19 and 42?	40 and 65?
7. 71 and 13?	17 and 28?	34 and 37?
8. 45 and 54?	45 and 19?	19 and 46?
9. 11 and 75?	57 and 13?	41 and 29?
10. 15 and 51?	81 and 17?	22 and 44?
11. 37 and 45?	66 and 24?	44 and 28?
12. 47 and 54?	23 and 32?	55 and 39?
13. 39 and 45?	45 and 57?	64 and 47?
14. 93 and 25?	48 and 56?	76 and 58?
15. 19 and 91?	51 and 62?	79 and 69?
16. 46 and 52?	59 and 73?	88 and 77?

Note.—Similar questions may be carried to any desirable extent. A single exercise, however, ought not to be continued over 15 minutes.

CHAPTER XII.

§ 84.—Analysis, in Arithmetic, as the thoughtful student has already been able to infer, is the process of ascertaining the operations to be performed in the solution of questions, unaided by specific rules.

§ 85.—Analysis may be applied in particular examples, for the purpose of deducing general principles and rules; and also to the direct and simple solution of a great variety of questions, both in common arithmetic and the practical doings of life.

EXAMPLES ANALYZED.

1. If 15 barrels of pork cost $195, how much will 7 barrels cost?

ANALYSIS.—1 is $\frac{1}{15}$ of 15; hence 1 barrel will cost $\frac{1}{15}$ as much as 15 barrels; $\frac{1}{15}$ of $195 is $13. If 1 barrel cost $13, 7 barrels would cost 7 times $13; 7 times $13 are $91. Therefore, if 15 barrels of pork cost $195, 7 barrels would cost $91.

2. If 7 barrels of flour cost $30, how much will 9 barrels cost?

3. If 12 barrels of beef cost $75, what will 7 barrels cost?

4. A drover bought a horse and paid $16 down, which was $\frac{2}{5}$ of the price of it; what was the price of the horse?

ANALYSIS.—From the conditions of the question it is plain, that $16 is $\frac{2}{5}$ of the price of the horse. If $16 is $\frac{2}{5}$, $\frac{1}{2}$ of $16 is $\frac{1}{5}$ of the price; $\frac{1}{2}$ of $16 is $8. If $8 is $\frac{1}{5}$, 5 times $8 are $\frac{5}{5}$ or the whole price; 5 times $8 are $40. Therefore, the price of the horse was $40.

Note.—In solutions of this character it is difficult for some pupils to see that $\frac{1}{5}$ is $\frac{1}{2}$ of $16, &c. They reason that if $16 is $\frac{2}{5}$, $\frac{1}{5}$ is $\frac{1}{5}$ of $16, &c. Pupils will discover the error of this mode by having them substitute the word ***parts*** in the place of the denominator.

5. A man bought 4 hundred weight of yellow ochre for \$48; what would have been the cost of 7 hundred weight, at the same rate?

6. 144 is $\frac{12}{13}$ of what number?

7. $\frac{4}{9}$ of 36 is how many times 5?

ANALYSIS.—$\frac{1}{9}$ of 36 is 4; $\frac{4}{9}$ is 4 times 4; 4 times 4 is 16. As many times 5 as the number of times that 5 is contained in 16; 5 in 16, $3\frac{1}{5}$ times. Therefore, $\frac{4}{9}$ of 36 is $3\frac{1}{5}$ times 5.

8. A man paid \$96, which was $\frac{5}{8}$ of all the money he had, for flour at \$6 a barrel; at the same rate per barrel, how many barrels could he have purchased with all his money?

9. A farmer had \$40, and paid $\frac{5}{7}$ of it for 14 barrels of cider; what was that per barrel?

10. A farmer bought a yoke of oxen and paid \$50 down, which was $\frac{7}{10}$ of the price; how many cords of wood, at \$3 a cord, will it take to pay the balance?

ANALYSIS.—If \$50 is $\frac{7}{10}$, $\frac{1}{7}$ of \$50 is $\frac{1}{10}$; $\frac{1}{7}$ of \$50 is $\$7\frac{1}{7}$; and 10 times $\$7\frac{1}{7}$ are $\frac{10}{10}$. 10 times $\$7\frac{1}{7}$ are $\$71\frac{3}{7}$, the price of the oxen. Having paid \$50, he still owes the difference between $\$71\frac{3}{7}$ and \$50. $\$71\frac{3}{7}$ less $\$50 = \$21\frac{3}{7} = \$\frac{150}{7}$. It will take as many cords of wood to cancel the debt as the number of times \$3 is contained in $\$21\frac{3}{7}$ or $\$\frac{150}{7}$. $\$3 = \$\frac{21}{7}$. Dropping the denominators we have $\$\frac{150}{21}$ or $\$150 \div \$21 = 7\frac{1}{7}$. Therefore it would take $7\frac{1}{7}$ cords of wood to pay the balance of the debt. Or as many cords as the number of times that \$3 is contained in $\$21\frac{3}{7}$, which is $7\frac{1}{7}$ times.

11. A man sold 42 sheep, which was $\frac{5}{7}$ of his entire flock; he afterward sold the remainder at \$2 per head; what did they come to?

12. 108 is $\frac{9}{11}$ of how many times 8?

13. $\frac{5}{8}$ of 40 is $\frac{2}{7}$ of what number?

ANALYSIS.—$\frac{1}{8}$ of 40 is 5; $\frac{5}{8}$ is 5 times 5, or 25. If 25 is $\frac{2}{7}$ of some number, $\frac{1}{2}$ of 25 is $\frac{1}{7}$ of the same number. $\frac{1}{2}$ of 25 is $12\frac{1}{2}$. If $12\frac{1}{2}$ is $\frac{1}{7}$, 7 times $12\frac{1}{2}$ is $\frac{7}{7}$; 7 times 12 is 84, and 7 times $\frac{1}{2}$ is $\frac{7}{2} = 3\frac{1}{2}$, which added to $84 = 87\frac{1}{2}$. Therefore, $\frac{5}{8}$ of 40 is $\frac{2}{7}$ of $87\frac{1}{2}$.

14. $\frac{4}{5}$ of 30 is $\frac{2}{3}$ of what number?

15. $\frac{7}{9}$ of 81 is $\frac{7}{15}$ of what number?

16. If 40 pounds of sugar cost $3.20, how much will 9 barrels cost?

17. If 7 tons of hay cost $31, what will 8 tons cost?

18. If 6 men can do a job of work in 13 days, in what time can 11 men do it?

19. If 12 pounds of coffee will last a family of 5 persons 8 weeks; how long will it last them if 3 persons are added to the family?

20. If 7 men can dig a ditch in 13 days, in what time can 11 men dig one 3 times as long?

21. When potatoes are 50 cents a bushel, how many bushels must I give in exchange for 18 bushels of wheat, at 75 cents a bushel?

22. How many yards of carpeting that is $\frac{3}{4}$ of a yard wide, are equal to 17 yards that is $\frac{7}{8}$ of a yard wide?

23. What number is that to which, if its third and 64 more be added, the amount will be double the number?

24. If $\frac{1}{5}$ of a pole stand in the water, $\frac{1}{4}$ in the mud, and 10 feet above the water, what is the length of the pole?

$\frac{1}{5}$ and $\frac{1}{4}$ are equal to $\frac{9}{20}$; hence $\frac{11}{20} = 10$ ft., the part above the water.

25. A messenger traveling at the rate of 6 miles an hour, was sent to Mexico with dispatches for the American army; after he had been gone 7 hours, an-

other was sent with countermanding orders, who could travel 15 miles in the same time that the first could go 12; how long would it require for the latter to overtake the former; and how far must he travel?

SOLUTION.—From the conditions of the question, it appears that the second gained on the first $1\frac{1}{2}$ miles per hour; and will, therefore, be as many hours in overtaking him, as the number of times that $1\frac{1}{2}$ is contained in 7 times 6 or 42, namely, 28 hours: and 28 times $7\frac{1}{2}$m. = 210m., the distance traveled.

26. The head of a certain fish was found to be 9 inches long; its tail was as long as its head and $\frac{1}{2}$ of its body, and the body was as long as the head and tail both; what was the length of the body?

SOLUTION.—Since the body was as long as the head and tail, it follows that it was $\frac{1}{2}$ the length of the fish, and since the tail equals the head and $\frac{1}{2}$ the body, it must have been 9 in. plus $\frac{1}{4}$ of the fish.

If now we draw a line representing the fish, we shall readily discover his length.

Head.	Body.	Tail.
9 in.	$\frac{1}{2}$	9 in. + $\frac{1}{4}$.

Here we see that $\frac{1}{2} + \frac{1}{4} = \frac{3}{4}$ and 18 in. more = the length of the fish. If then 18 in. is $\frac{1}{4}$ the length, $\frac{2}{4}$ or $\frac{1}{2}$ of the fish is 2 times 18 inches or 36 inches, which is equal to 3 feet.

27. A hare starts 50 leaps before a greyhound, and takes 4 leaps to the hound's 3; but 2 of the hound's leaps are equal to 3 of the hare's; how many leaps must the hound take to overtake the hare?

SOLUTION.—Since 2 of the hound's leaps = 3 of the hare's, 1 of the hound's = $1\frac{1}{2}$ of the hare's. Hence, while the hare is taking 4 leaps, the hound advances

3 times $1\frac{1}{2}$ of the hare's leaps. $1\frac{1}{2} \times 3 = 4\frac{1}{2}$; therefore, in taking 3 leaps, the hound gains $\frac{1}{2}$ of the hare's leaps, and must take 6 leaps to gain 1 of the hare's leaps; and to gain the 50 leaps, he must take 50 times $6 = 300$ leaps.

28. A workman was engaged for 40 days upon this condition, that he should receive 20 cents for every day he wrought, and should forfeit 10 cents for every day he was idle; at settlement he received 5 dollars; how many days did he work, and how many days was he idle?

29. A, B, and C, bought a quantity of wine for \$340, of which sum A paid three times more than B, and B four times more than C; how much did each pay?

30. A man and his wife can eat a barrel of beef in 15 weeks, but after eating 9 weeks, it was found that his wife could eat what remained in 30 weeks; in what time could either have eaten the whole?

31. A young man having a patrimony of \$36,000, spent $\frac{3}{4}$ of it in gambling and dissipation; how many acres of land, at \$5 an acre, could he buy with the remainder, after paying \$500 for farming implements?

32. If a yard of silk cost 50 cents, what will 360 yards cost?

Note.—The pupil will note that 50 cents is equal to \$$\frac{1}{2}$.

33. A pillar of granite stands 40 feet above the surface of the snow, $\frac{1}{5}$ in the snow, and $\frac{1}{16}$ in the ground; what is the whole length of the pillar?

34. What number is that to which, if $\frac{1}{5}$ and $\frac{1}{6}$ of itself be added, the sum will be 164?

35. If 12 horses consume 720 bushels of oats in 3 months, in what time will 18 horses consume 1200 bushels?

36. If 10 horses consume 20 tons of hay in 30 weeks, what number of horses will it require to consume 40 tons in 15 weeks?

37. 36 barrels of flour will last 840 men 5 days; how many barrels will last 1200 men 11 days?

38. A returned Californian being asked how much money he had, answered that $\frac{2}{3}$, $\frac{3}{4}$, and $\frac{5}{8}$ of it made \$147,000; what amount had he?

39. What number is that, $\frac{2}{3}$, $\frac{3}{4}$ and $\frac{1}{12}$ of which is 390?

40. What number is that, $\frac{7}{8}$ of which exceeds $\frac{4}{5}$ of it by 18?

41. What number is that to which, if its $\frac{1}{2}$ and $\frac{1}{3}$ be added, the sum will be equal to 5 times 11?

42. A farmer being asked how many sheep he had, replied that, if he had as many more, $\frac{1}{2}$ as many more, and $2\frac{1}{2}$ sheep, he should have 100; what number of sheep had he?

43. A clothier bought $8\frac{1}{3}$ pounds of indigo for \$$20\frac{5}{6}$; what would have been the cost of $5\frac{1}{7}$ pounds, at the same rate?

44. The distance from A to B, which is 40 miles, is $\frac{4}{5}$ of $\frac{1}{2}$ the distance from C to D; what is the distance from C to D?

45. A has $\frac{3}{4}$ as much money as B, and $\frac{4}{5}$ as much as C, who has $\frac{5}{6}$ as much as D, and he has \$1800. What are the respective sums owned by A, B, and C?

CHAPTER XIII.

§ 86.—*Note.*—In this chapter the same principles are used as in previous sections, the chief difference being, that larger numbers are used. No mental exercise is of more value than that of adding large numbers mentally. The books of the class being closed, let the following and similar examples be much dwelt upon. The teacher should solve the questions, as well as the pupils. The practice will be of value, and will also furnish a guide, as to the rapidity with which the numbers should be read. Let all remain silent until the final result is secured.

1. What is the sum of 18, 13, 19, 15, 17, and 14?
2. What is the sum of 16, 20, 29, 25, 24, and 27?
3. What is the sum of 23, 28, 32, 37, 38, 41 and 49?
4. What is the sum of 45, 53, 44, 56, 48, 47 and 58?
5. What is the sum of 38, 48, 58, 83, 99, 75 and 98?

Arithmetical Signs.

§ 87. The sign of ADDITION consists of two lines, one horizontal and the other perpendicular, thus, +, and shows that the numbers between which it is placed, are to be added together. It is usually read *plus*. Thus the expression 4 + 7, is read 4 *plus* 7.

§ 88. The sign of EQUALITY consists of two parallel horizontal lines, thus =, and shows the numbers or quantities between which it is placed, are equal to each other.

Thus 4 + 7 = 11, is read 4 *plus* 7 equal 11.

§ 89. The sign of MULTIPLICATION is commonly denoted by two oblique lines crossing each other, thus ×, and shows that the numbers between which it is placed, are to be multiplied together. Thus 4 × 7 = 28, is read 4 multiplied by 7 = 28. A dot or point placed between the numbers, also signifies the same thing. Thus 4 . 7 signifies the same as 4 × 7, and is read in the same manner.

§ 90. The sign of SUBTRACTION is represented by a short horizontal line, thus —, and is read *minus*.

When placed between two numbers it denotes that the number before it is to be diminished by the number after it. Thus 7 — 4, shows that 7 is to be diminished by 4, and is read 7 minus 4, or 7 less 4.

§ **91.** The principal sign of DIVISION is a short horizontal line placed between two dots, thus ÷, and shows that the number before it is to be divided by the number after it. Thus 24 ÷ 6 shows that 24 is to be divided by 6, and is read 24 divided by six. Division is also, very often, expressed by writing the divisor or number to divide by, under the dividend or number to be divided, in the form of a fraction, thus $\frac{24}{6}$; and signifies the same as 24 ÷ 6, and is read in the same manner.

§ **92.** A line, called a *vinculum* ———, or a parenthesis ( ), is used to denote that the same operation is to be performed on two or more numbers. Thus $\overline{3 + 7} \times 5$, or $(3 + 7) \times 5$, shows that the sum of 3 and 7 is to be multiplied by 5. If the expression was $3 + 7 \times 5$ it would denote that 7 only was to be multiplied by 5, and the product added to 3.

§ **93.** Example 1. What whole number is equivalent to the following: $3 + 18 \times 11 - 54 \div 3 + \frac{12}{4} - 5 \times 9 + 1 + (3 + 4) \times 7$?

2. $5 \times 6 \div 15 - 1 + 35 \div 6 \times 4 + 12 + 16 - 8 + \overline{3 - 1} + 14 \div 5 \times 4 + \frac{14}{7}$ = how many?

3. $13 \times 15 + 24 \div 4 - 18 - 10 + 20\frac{1}{4} \div 11 + 3 \times 12 + 18 + 7 - 4 + 24 + 36$ = how many?

4. $14 - 3 + 33 \times 5 + 88 \times 2 + 60 - 111 + \frac{96}{8} - (3 \times 4) + 18 - 11 + 222$ = how many?

5. What is the difference between $36 \div 9 + 17$ and $36 \times 9 - 17$?

6. What is the sum of 95, 84, 73, 62, 51, 40, 39, 28, and 17?

7. State the sum of the following numbers, viz: 13, 24, 35, 46, 57, 68, 79, 80, and 91.

8. What is the sum of all the numbers from 1 to 60, inclusive?

9. What is the difference between all the numbers from 40 to 50, inclusive, and 50 to 60, inclusive?

10. What is the sum of all the numbers from 1 to 100, inclusive?

PART II.

TO TEACHERS—MODE OF CONDUCTING CLASSES.

NOTHING is more ungenerous in an author, than to fix arbitrary modes of solving any difficulties that may be interwoven into the subject being treated; for no one has a right to resort to any mode of demonstrating truths, that he does not clearly comprehend.

A number of excellent modes of teaching *Intellectual Arithmetic* are in almost daily use among a large number of teachers, and some few have been given to the world. It is no part of my design to dictate any prescribed rules at this time for teaching this science, but simply to point to a few of the most common errors that have fallen under my observation, which every intelligent teacher will seek to avoid; and with a few general hints on *better modes*, leave the subject with you to modify and improve as circumstances shall seem to dictate.

Errors in Teaching Mental Arithmetic.

1*st Error.*—The practice of allowing each scholar to pursue the study on his "own hook," without giving recitations, and receiving no instruction from his teacher, save now and then the solution of a problem, without knowing the "why and wherefore" of a single point in any solution, is a prominent evil with many instructors. This mode of teaching, or rather want of teaching, has been justly reckoned as belonging to days gone by; and, says Prof. Thomson, "it is *prima facie* evidence that those who practise it, are *behind the spirit* of the times."

2*nd.*—Another error is, allowing pupils to use the slate and pencil, which are emphatically not the instruments for procuring a knowledge of *Intellectual* Arithmetic. Teachers allowing this error a place, are worthy only of being likened to the first transgressors.

3*rd.*—A disposition to discourage inquiry on the part of pupils, is another crying evil that many teachers are guilty of. That we may be forgiven many sins is true; whether he who breaks down the rising energies of an inquiring, youthful mind, is a question.

These are perhaps the most glaring errors, that need a rebuke in this place. We pass to notice a very few errors in the language of solutions. Avoid such expressions as the following, viz:

1. 3 times 15 are 45, 4 times 6 are 24, &c. The verb in these and similar cases is *singular*, and should have the singular form. (*See Bullion's Analytical and Practical Grammar*, page 39; also *Webb's 2nd Reader*, page 20.)

2. *Three is the one third* of three times itself; for, three is one third of three times three.

3. In questions like those to be found in Sec. 22 of this work, (example 22 for instance,) most teachers allow the question to be repeated in the midst of the solution. This not only weakens the power of the problem in promoting *mental discipline*, but at the same time mars the beauty of the solution. (*See Sec.* 22, *Example* 22, Solution.)

4. Many teachers, desirous of having their pupils "show off" well, are in the habit of drawing the solutions of the questions out of them by littles, taking care to tell the pupil the answer to the question before asking it. This practice cannot be too strongly condemned. For an illustration of this mode of teaching, see "*Page's Theory and Practice of Teaching,*" under head of "Drawing out Process."

5. A fifth error is that of passing over the questions, requiring the answer simply, without a single reason for it—without the least shadow of a demonstration.

6. The sixth and last error we shall give in this connexion, is that of allowing pupils to leave one subject and pass to another, before the first is understood and learned. The practice of this error has probably made more superficial students, than all the others combined. Parents and teachers cannot too carefully avoid it.

Suggestions on the Best Mode of Teaching Arithmetic.

In order for a person to become a successful teacher of Arithmetic, as well as any other branch of science, he must have a *thorough knowledge*, both of the *general principles*, and of the different modes of illustrating those principles to the young. To secure both of these means, he must be possessed of a natural love for the branch.

Pupils of nearly the same degree of advancement, should be formed into a class—and regular lessons assigned, to be recited each day at some particular hour. The advantages of having a school properly classified, are numerous and important.

1. It is a great saving of time to the teacher.

2. A more powerful stimulant to exertion can be awakened, and explanations given to the whole class as easily, as to each individual separately.

3. As much, and oftentimes much more, is learned in thirty minutes on the recitation seat, as could be learned by the pupil, apart from such recitation, in whole hours, and perhaps days. By hearing the different modes of solution, the teacher himself often gets new modes of illustration, which may be of great value to him in the future practice of his profession.

The prime object in every recitation, is to bring something to the minds of the pupils that they have not found in the book, and which will not only give the class increased confidence in the teacher, but will at the same time secure the attention of every member of the class.

Every question and example should be fully and clearly analyzed, the reason for every step explained, until such time as each scholar is able to go through with the analysis without difficulty or mistake.

Whatever may be the condition of a school, the motto of all teachers should be *perfection;* and unless this motto is lived up to, they must unavoidably fail of success.

Every pupil, however young, should be fully impressed with the idea that *he must depend upon himself.* He *must* do his own *learning.* Whenever it becomes necessary to render any scholar assistance, it should be done indirectly, and not directly. The teacher's business is not *to do* the work of the pupil, but to *teach him how to do it.*

Books in Recitation.

In the recitations, the pupils should generally be allowed to have their books open before them, the readiness and accuracy with which they perform the examples, being made the test of qualification.

Modes of Solution.

The forms of solution given, are not designed to be used in all cases, but may at any time be discarded if a *better* mode can be given to supply its place. The language of the solutions used should, in general, be uniform, in the same school. A pupil should in no case be allowed a form of solution that he does not *himself* clearly *comprehend.*

The tables should be recited with the books closed. If more questions are desirable than are contained in the several sections, they can be added under each head by the teacher.

Note.—A smaller work for children is in process of preparation, and will be ready for the press soon; which, it is hoped, taken in connexion with this, will render pupils who pursue them systematically and thoroughly, fully prepared to enter the study of *written arithmetic,* and gain a speedy and extensive knowledge of it without difficulty.

An Elementary and Practical Arithmetic, in which have been made VARIOUS IMPROVEMENTS in arrangement and nomenclature, as well as the means of securing THOROUGH DISCIPLINE in the principles and applications of the science, by James B. Dodd, A. M., of Transylvania University, Ky., is designed to follow this work.

www.ingramcontent.com/pod-product-compliance
Lightning Source LLC
LaVergne TN
LVHW021410110826
845150LV00007B/1867